Apollo Root Cause Analysis

A New Way Of Thinking

By Dean L. Gano

Third Edition
Second Printing

Apollonian Publications, LLC
Richland, Washington

Apollo Root Cause Analysis — A New Way of Thinking

Call Atlas Books for Additional Copies; 800-247-6553

ISBN 1-883677-11-4

Copyright Registration Number TX5-432-060

*To my students, who have
taught me well*

Preface

A New Edition Driven by Continuous Learning

In the first edition of *Apollo Root Cause Analysis – A New Way of Thinking*, published in 1999, I expressed the hope that "this book would stimulate more learning for me and others who are interested in this fascinating subject." I am happy to report my hopes are being realized because Apollo Root Cause Analysis (ARCA) is steadily becoming the world standard for event-based problem solving. Since the late 1980s, more than 75,000 people have been trained world-wide, and more than 50,000 books have been sold. The training material has been translated into 9 languages and is used by businesses of all sizes, from the largest to the smallest. It is used by professionals, crafts, managers, and workers of all kinds in a large variety of industries from Aerospace and Information Technology to Healthcare and Agriculture. Indeed, it can be used on any problem where we interact with our environment. With this increased exposure and more minds contemplating the use of the methodology, we have all learned even more; and this new edition reflects what we have learned in the last 8 years.

Perhaps the most important thing we have learned is the powerful effect RealityCharting® software has on problem solving. By providing an intuitive communications tool based on the proven ARCA methodology, RealityCharting® facilitates a much more robust cause and effect chart—called a Realitychart (The name "Realitychart" was chosen because the software creates a common reality based on the input of all stakeholders). This updated edition of *Apollo Root Cause Analysis – A New Way of Thinking* includes helpful information on how to use the software to maximize your problem-solving effectiveness.

Table of Contents

Introduction

The purpose of this book is to share what I have learned about effective problem solving by exposing the ineffectiveness of conventional wisdom and presenting a principle-based alternative called Apollo Root Cause Analysis that is robust yet familiar and easy to understand. I hope this book will improve your understanding of human problem solving and thus make your journey in life more successful.

This book will change the way you understand the world without changing your mind. One of the most common responses I get from students of ARCA is that they have always thought this way, but did not know how to express it. Other students have reported a phenomenon where this material fundamentally "re-wires" their thinking, leading to a deeply profound understanding of our world. Many ARCA practitioners have become the best problem-solvers in the company. One such person received a $50,000 bonus and recognition as the employee of the year. It is not unusual to see a return on investment of 10 times more than the cost of the event being investigated. An executive from one very large global company credits ARCA with saving it from bankruptcy by not only solving a multimillion dollar regulatory problem but by adding millions of dollars to the bottom line.

At the heart of this book is a new way of communicating that is revolutionizing the way people all around the world think, communicate, and make decisions together. Imagine your next decision-making meeting where everyone is in agreement with the causes of the problem and the effectiveness of the proposed corrective actions—no conflicts, arguments, or power politics! This is the promise of ARCA.

A New Way of Thinking

This book presents a unique and new way of thinking about the way we solve problems in our daily life. It is not about math problems or design problems. It is not about creative problem-solving or brainstorming. It is not another categorization scheme to lump causes and solve them by popular vote. It is not another compilation of conventional wisdom. Rather, it is about finding effective solutions to event-based problems—problems encountered when we interact with our environment.

The ARCA method is based on fundamental principles, which provide a standard or law that is applicable to every event-type problem. As such, it does not require that you have great subject-matter expertise. Simply by applying the fundamental principles discussed herein and having access to the subject-matter experts or the people who know the most about the actual causes of the event (usually the workers), anyone can solve any event-based problem every time. Furthermore, by understanding the cause and effect principle (Chapter 2), you will be more prepared to prevent problems in the first place—not only because you will begin to think causally and see problems before they occur, but with the help of RealityCharting® software you can create basic fault-tree diagrams that depict possible failure scenarios in your more complex systems.

Effective Problem Solving

In everything we do, the ability to solve problems effectively is fundamental to our success. The better we understand the underlying causes of an event, the better we can control it. The more causal relationships we understand, the better we can predict the future and the more successful we will become.

While we inherently understand the notion of cause and effect, no culture has ever devised a principium of cause, and hence we have never taught causal strategies in our schools. The paradigms and mental models on which conventional wisdom are based tend to regard cause and effect as obvious. But in actuality, once you understand the complexities of the cause and effect principle (Chapter 2), "reality" becomes so complicated it boggles the mind.

The cause and effect diagramming process discussed herein has historically been called Apollo Root Cause Analysis, and the software application developed for it is called RealityCharting®. This simple and easy to use problem-solving method is based on the four elements of the cause and effect principle:

1. Cause and effect are the same thing.
2. Causes and effects are part of an infinite continuum.
3. Every effect has at least two causes in the form of actions and conditions.
4. An effect exists only if its causes exist at the same point in time and space.

Don't try to digest these elements right now, just make note that there are four laws that help define your reality. Applying these principles means that every time we ask "why" we must find at least two causes (third principle) and because cause and effect are the same thing (first principle), we must then ask "why" again. Because each effect reveals at least two causes (usually many more), each of those two causes must reveal two more for a minimum of four more, and those four become a minimum of eight and eight becomes 16, 32, 64, etc., on to infinity (second principle). The fourth element will be discussed in more detail later in the book, but we can see at this point that asking "why?" leads to an ever-expanding set of causes, something like the branches of a tree, limited only by our knowledge of the subject or event.

By understanding this complexity of causation we can now see why the human brain cannot deal with "reality." How could we possibly be able to comprehend, let alone communicate to others, an infinite set of causes for a single event? At every turn, our biological and cultural development has failed to devise a strategy to deal with this enormous complexity. Instead of dealing with this complexity, we have invented many simple-minded strategies to accommodate our simple minds and sadly to protect our fragile egos.

By understanding the cause and effect principle and creating a Realitychart, your understanding of what constitutes reality will be changed forever, and that change will allow you to become a more effective problem-solver. ARCA provides the methodology and RealityCharting® provides the tool to allow you to see a reality that was previously beyond your comprehension.

Beyond Conventional Wisdom of Problem Solving

This book exposes the fallacies of our most common problem-solving strategies and, more importantly, provides an alternative that works on every event-based problem every time. The most common problem-solving strategy in use today is to categorize causes or identify causal factors, look for root causes within these categories, and then vote on which ones we think are the root causes. There are many categorical schemes being used, like MORT, Ishikawa's Fishbone, and various "Cause Tree's" that come in an infinite variety. While these categorical methodologies claim to represent a causal analysis, there is no causal relationship whatsoever—simply groupings of cause categories based on the creators' own experiences within the system or world in which their events occur.

These categorical methodologies do help generate impressive Pareto charts, which can be used to prioritize which problems to work on, but they do not reliably lead to finding effective solutions. This is because cause categorization schemes do not reveal the cause and effect relationships crucial in finding effective solutions. Grouping or categorizing causes makes us feel good because we have taken what looks like a complex problem and simplified it into neat little piles of similar causes. True cause and effect analysis does the opposite. It takes what appears to be a few simple causes and humiliates us by revealing the complexity of what happened. The humility is worth it because of the richness and diversity of solutions that are revealed.

Unfortunately these "Cause Trees" are welcomed and used by many because they seem familiar and are easy to use. They are only familiar because they utilize biologically evolved processes of categorization inherent in our mental operating system. By design, our mind categorizes everything; it even stores nouns and verbs in different parts of the brain[1]. We will discuss categorization and other failed strategies further in Chapter 1 and in the Appendix.

Regardless of the problem-solving method(s) you currently use or don't use, you will find Apollo Root Cause Analysis and the companion RealityCharting® software a simple, structured, and more powerful alternative. After you have read this book, I invite you to visit our Web site at www.realitycharting.com and begin using RealityCharting® immediately at no cost without having to log in or register. Use it to create a Realitychart of any event you have interest in, or visit RealityCharting > Example Problems; URL http://www.realitycharting.com/realitycharting/example-problems/ to see a variety of Realitycharts, including one analyzing ineffective problem solving. RealityCharting® is simple and intuitive—watch the 2-minute tutorial and then get started using a tool that will significantly improve the way you and those around you understand the world.

The ARCA process eliminates the usual arguing and politics associated with most event investigations, and results in effective solutions that all stakeholders can buy into. While our individual realities will always remain unique, we can better understand the realities of others and share ours with them to form a common reality through the application of the cause and effect principle and the use of RealityCharting®.

How To Read This Book

I have designed *Apollo Root Cause Analysis – A New Way of Thinking* to be read at several levels and to allow quick reference to topics or sections. Because this book challenges conventional wisdom, it may not validate your existing belief system, so I suggest you read the book straight through first and then go back and reread the chapters that interest you the most. Highlight, underline, and otherwise dog-ear the book as you go so that it becomes a useful reference for you. Many very important messages are contained in single sentences throughout the book, and you will want to note them and carefully consider their value.

Chapter 1 examines the typical ways most of us approach problem solving, where those strategies came from, and why those strategies rarely keep our problems from recurring. Chapter 2 describes the cause and effect principle, upon which the Apollo method is based. Chapters 3, 4, and 5 describe the ARCA tools that are critical for implementing the cause and effect principle and finding effective solutions. These chapters provide "how-to" information with some guidelines and philosophy interspersed. Chapter 6 provides valuable guidance on facilitating a team using the ARCA method. The ARCA method can be used individually or in a group and is most powerful in a group. Chapter 7 highlights the attributes of the Apollo method.

In the Appendix, I have summarized, for comparison, conventional root cause analysis methods.

Finally, the Glossary provides a single-source reference for the new words you will learn, and the Index helps you get to specific topical discussions.

Training

While the Apollo process is simple and the software intuitive, we have found that most people benefit significantly from a formal training program. Training is available from certified ARCA/RealityCharting® instructors. A 2-day course helps to further reinforce these concepts through insightful and practical exercises, and graduates receive a copy of the RealityCharting® software and certification as an ARCA practitioner and team facilitator. A 1-day course is also available for management and others who may need practical knowledge, but who won't be facilitating any investigations. If you are interested in training, software, or other services, please see the contact information provided at www.realitycharting.com.

About the Author

Dean L. Gano is President of Apollonian Publications, LLC, which is dedicated to helping others become the best event-based problem-solvers they can be by providing highly effective problem-solving tools in the form of books and computer software. Mr. Gano brings more than 36 years of experience in process industries, power plants, and computer software development to this endeavor. He started his incident investigation work and subsequent fascination with problem solving while working on solutions to the incident at Three-Mile Island Nuclear Power Station in the late 1970s.

He has participated in hundreds of incident investigations since and studied the problem-solving process through his teaching and consulting work. He has been teaching his unique version of root cause analysis to people around the world for more than 20 years. His message is being taught in nine different languages on five continents and is being used globally by many of the Fortune 500 companies, the National Aeronautics and Space Administration, the Federal Aviation Administration, and other government entities. To ensure an effective problem analysis is done every time, Mr. Gano created RealityCharting® software, which is becoming the world standard problem-solving tool.

Mr. Gano holds Bachelor of Science degrees in Mechanical Engineering and General Science, is a formerly certified nuclear reactor operator, and a Vietnam veteran. He is a senior member of the American Society for Quality and the American Society of Safety Engineers. He is a philosopher and student who finds great happiness in learning and helping others become more successful in their life pursuits.

Reference

1. Carter, Rita 1999 *Mapping The Mind,* University of California Press, London, UK

1
Set Up To Fail

Ignorance is a most wonderful thing.
It facilitates magic.
It allows the masses to be led.
It provides answers when there are none.
It allows happiness in the presence of danger.

All this, while the pursuit of knowledge can only destroy the illusion. Is it any wonder mankind chooses ignorance?

In every human endeavor, a critical component to our success is our ability to solve problems. Unfortunately, we often set ourselves up to fail with our various problem-solving strategies and our inherent prejudices. We've typically relied on what we believe to be common sense, storytelling, and categorizing to resolve our problems. Conventional wisdom has us believe that problem-solving is inherent to the subject at hand—the doctor solves medical problems, the mechanic fixes our car, etc. Using these strategies often leads to conformity, which brings complacency and mediocrity. This chapter will expose the failed strategies that prevent us from being effective problem solvers.

As we explore the reasons behind ineffective problem solving, we will see how rule-based thinking creates the illusion of one right answer and a misguided belief in common sense. We will also see how our natural prejudices prevent effective problem solving. By dispelling the notion of common sense, we are able to replace it with a common reality that allows extremely effective communication. By appreciating all views and seeking causes, not blame, we will start down a path that leads to effective solutions for everyday problems every time.

Please open your mind and join me on an adventure into a new way of thinking that will improve your problem-solving skills and enable much more effective communications.

Problem solving is generally understood to mean overcoming some kind of difficulty by implementing a solution. The best solutions are often the most difficult to find, not because they are hiding but because we don't know how to find them. We call these untapped solutions creative solutions because they are seemingly created from inside our minds. Like the sculptor's notion that the statue lies within the stone, many effective solutions are waiting to be revealed by the method provided herein.

Depending on the various abilities we start out with, combined with the experiences we encounter in life, we each develop our own strategies for coping with life's problems. We each define our own world by creating our own reality. We observe how different things interact and establish our own understanding of the world through these relationships. We learn to control various causes (for example, people and things) to obtain certain goals. We do all this without even knowing it. It is simply part of our nature to explore and understand the world around us. Problem-solving skills vary significantly from person to person, and unfortunately most are ineffective.

The notion of a single right answer, the belief in something we call "common sense," and the natural tendency to establish biases and prejudices are all strategies that block effective solutions. This chapter will explore some of these strategies and where they come from.

Problem Solving

One of the most difficult questions I am asked as I travel around the world is, "What do you do for a living?" The answer to this question is difficult because most people are not familiar with what I do. My usual response, "I teach people how to be better problem solvers," is understood in many ways, but rarely as I intended. "Oh, are you a psychologist?" "Oh, are you a college professor?" "What kind of problems?" "Are you a management consultant?" I am frustrated because I can't think of any other way to summarize what I do that will be understood. Every response reminds me how each person perceives the world differently and how the notion of problem solving has no common meaning.

Pondering why this is so, I wonder if it may be because our education systems do not recognize problem solving as an entity unto itself. Since problem solving has never been established as a separate subject or curriculum, our skills are not well developed. Aside from the sometimes boring and difficult dictums found in college courses on logic and critical thinking, no fundamental principles have been laid down on which to build a problem-solving curriculum. Problem solving is understood to be inherent in each subject, so problem solving for the computer engineer or the mechanic is thought to be unique to their occupations. Based on this belief, we have failed to teach effective problem solving.

I have discovered that while specific knowledge lies within the job, profession, or subject matter, effective problem solving can be universal to all subjects. Certainly most mechanics can't solve highly technical computer problems nor can the typical computer engineer be expected to rebuild an engine, but they both can use the same problem-solving strategies in their work and their lives.

While problem solving can be categorized in many ways, we usually treat problems as if they are rule based. That is, we seem to believe all problems have "one right answer." A colloquial saying even expresses this notion: "It's the right thing to do." Many people are so intent on solving all problems with rules that they limit themselves to the same old favorite solutions that failed to prevent them from recurring in the first place.

Rule-based problems follow rules created by people to help us understand repeatable events, such as a company procedure or established laws. In rule-based problems we agree to a convention, and thus a single answer or pre-defined solution is usually available, for example, $2 + 2 = 4$, or if we run a red light we may be fined, or score the most points in a game and you win. In each case, the answer is predefined by a set of rules that all players agree to. The rule-based approach is often more concerned with conformity and consistency than with accomplishing our goals. The rule-based approach to problem solving is often ineffective because our daily lives are filled with the immense variability of the human condition. As such, most problems do not have one right answer—only good, better, and best. These daily problems are called event-based problems; they are problems we encounter when interacting with our environment.

The concept of the "right answer" was brought home to me a few years back when I was teaching a class at a national laboratory. As we will discuss later, asking "why" is an important part of my approach in identifying causes and effects in problem solving. A Ph.D. physicist, who was also a tenured professor at a prestigious college, informed me during class that to ask "why" was foolish. He talked about Einstein and Niels Bohr and stated that "why" should never be asked. Since this was a total affront to the theme of my class, we had many discussions over the next two days and I finally came to understand his perspective. In his world of experimental physics, he always establishes a box around the experiment so there are no unknown variables. In scientific experiments, everything is known, a condition is changed, and the result is documented and used to provide evidence of a theory or premise.

When I discovered his perspective, I pointed out to him that the world outside his boxes did not have the luxury of complete knowledge. Variables exist in the infinitum, I explained. He understood what I had said, but it destroyed his illusion of the perfect world where everything is known and right answers always exist.

When we cling to a rule-based mindset, we set ourselves up to fail when trying to solve event-based problems in daily life. We so often look for one right answer because that is what we have been taught to do. The next section takes a look at the practices we have so carefully cultivated and refined but unfortunately allow us to repeat our problems rather than prevent them from happening.

Typical Problem-Solving Practices

When I first began teaching root cause analysis, I taught some of the conventional wisdom of the day. I taught people how to categorize causes and how to find the "real" root causes. In each class, a few students would seriously challenge what I was teaching. With an aversion to contradictions, I reflected on each class and was constantly learning and changing what I taught. In the process, I identified several problem-solving strategies that are detrimental to effective problem solving. The most common detrimental problem-solving practices used by individuals and organizations the world over include stopping too soon, the need to place blame, the root cause myth, the false belief in common sense and a single reality, groovenation, storytelling, and categorical thinking. We are going to examine each one of these failed strategies and as you will see many of them are interrelated.

Stopping Too Soon

For the most part, we have been set up to fail by a culture that has never adopted a principium of cause. Perhaps this is because we operate on the general assumption that everyone has basic problem-solving skills, and all they need bring to the table is their specific knowledge. Consider the following real example of an empowered work group at one of North America's larger manufacturers.

"Hey, Frank, our building needs more lights! Is it OK for us to order new lights? I mean management said they would support us, so we should be able to just do it, right?" Seeking to better understand the situation, Frank asked, "What do you mean you need more lights?"

"Well, in our quality circle the other day, the guys decided the lighting in here is no good. When we asked management for money to buy more lights, they wouldn't give us any.

Again, Frank persisted, "So tell me more about your lighting problem. I hear a problem and a solution in the same sentence. Why do you think the lighting is so bad?"

"Like I told you, there ain't enough lights in here."

"Well, can you show me what you mean?" Frank continued.

As Frank looked around the building for possible causes, he found light fixtures covered with several layers of white paint. New cable trays had been installed and blocked the light. Several light bulbs were burned out, and one lighting circuit was not working. Frank had the fixtures cleaned or replaced and cable trays moved. No additional lights were needed.

Fortunately for this company they had Frank, who understood effective problem solving. The people in the quality circle had a solution without fully understanding the problem. As a result, their solution was inappropriate. If we are going to empower people, we need to make sure they have good problem-solving skills, or we are setting them up to fail.

A recent survey[1] documented just how bad our problem-solving skills are. The survey, which was a limited nonscientific study, revealed that only 20% of the general populace understand the concept of basic problem solving. Five simple events were presented to each participant. Each event had an unacceptable consequence, and the participants were asked to place themselves into the situation and find out what happened so they could prevent the problem from happening again. The entire conversation was recorded and documented.

While evaluating the responses is subjective, clear trends emerge. In about 10% of the responses, participants immediately sought to place blame. Another 26% immediately expressed an opinion of the causes and offered a solution without investigating the problem. It was encouraging to find 50% of the participants immediately asked "why," yet most stopped this line of questioning after only two or three "whys." When they stopped asking "why," their solutions either sought to place blame or was a favorite solution that seemed to fit.

Only two out of every ten responses continued to pursue causes until they found enough cause and effect relationships to allow an effective solution to be implemented. This is a significant indictment of the general populace's problem-solving skills. My own study of industry in the United States and abroad indicates that we only find effective solutions about 30% of the time. It doesn't matter whether it is a safety incident, equipment failure, or customer service issue. Regardless of the industry, the company, or the country, I have found that companies' incident reports reflect the same symptoms and the same poor problem-solving skills as the study discussed above. Furthermore, this failure goes beyond the incident reports to the techniques people use and the way people think about problem solving.

Stopping too soon seems to be caused by the need to get on with a solution, which we will discuss later, but I also suspect that most people know they are not good at analysis so they rely on their past experience and just wing it. The fact that we seek to place blame about 20% of the time is also very disturbing because it is rarely an effective solution.

The Need to Place Blame

A contractor employee was driving his backhoe through the construction site when his boom struck an overhead power line. The subsequent line break caused a power outage and disruption of work over a large area. Safety investigators were promptly dispatched, and root causes identified. The first root cause was personnel error, and the veteran backhoe driver was fired. Other minor causes were identified, but the emphasis was on personnel error.

This is only one of thousands of examples that happen daily in American businesses, where punishment is perceived as an effective corrective action. Consider the example. Since this was a veteran backhoe driver (who by the way had never had an accident before), how will firing him prevent recurrence? Moreover, who learned the most from this event? The backhoe driver, of course. Firing this driver is like sending an employee to an expensive training course and then firing him when he returns. In effect, this company probably replaced the most experienced person with someone who has no experience with overhead wires. They may have reestablished the same conditions they had before the event. They have done nothing to prevent recurrence, and they have set someone else up to fail in the future.

The belief that punishment will improve behavior in adults is not supported by any facts or studies. In fact, most of the time, punishment causes the exact opposite behavior. If we are unduly punished, we do not strive to do better. We are more likely to seek revenge or to give up. Since we perceive ourselves as mature adults, we do not appreciate being treated as children. This often causes childlike behavior, which is not a question of maturity or self discipline, but a human reaction. You cause me pain, I react. The rational, reasoning "self" may not come into play when dealing with hurt feelings and emotional pain.

More important than the ignorance of our actions is the causes behind them. In the work place, we place blame because we don't know what else to do. Like parents, most supervisors and mangers are not prepared for their job. If we have not developed a philosophy for certain situations, we are forced to draw on other life experiences. In the case of personnel error and punishment, we look to similar past experiences. We may find them in a parent-child relationship, a military experience, a teacher-student experience, a theological teaching, or the criminal justice system that we read about every day. Since the workplace is not a family, the military, school, church, or prison, none of these experiences provide an effective reference for dealing with personnel performance problems in the workplace. We have been set up to fail by our environment.

Using punishment to prevent problems is rarely effective. Unless you believe, based on some evidence, that punishment will prevent recurrence of your problem, don't do it!

I was recently informed that one of our clients wanted to modify the Apollo method to allow disciplinary action as a solution. To my great surprise, he thought the Apollo method does not allow discipline. It seems they wanted to punish their employees and couldn't do it with the Apollo method, so they asked us if they could change our method a little. Contrary to this perception, the Apollo method absolutely supports discipline, but only under circumstances where this is appropriate.

Discipline can be two different things. It can mean the punishment or the praise of an individual to effect a change in behavior. With punishment, the purpose is to stop undesired behavior. A person being praised understands they should continue their behavior. In this sense, discipline is an act of reinforcement and comes in one of two forms, positive or negative. It can be self-generated or come from outside.

When discipline comes from within, be it positive or negative, we accept it as having some value. When discipline comes from outside ourselves, it will seem appropriate and cause change only if we agree with it. To agree we must see the value in the discipline. While the value is obvious if it is positive reinforcement and allows us to meet our goals, it is something altogether different for punishment.

Value in punishment may be harder to accept, but is not unusual. Most people learn at a young age that if they violate established rules, punishment will surely follow. If taught and learned, we accept this causal relationship as a fact of life. Speeding on the freeway is a good example of this relationship. Not only do I know the speed limit as well as how fast I am going relative to it, I use a radar detector to reduce the likelihood of being caught. If caught, however, I accept the consequences and may modify my behavior. In the long run, I may even value this discipline as helping me grow to be a more responsible adult.

In the workplace we may violate established rules using the same logic. The thinking may go like this: "I know it's wrong, but as long as I can get away with it, I am more productive. I will accept the consequences if I get caught."

In general, we accept the consequences of our actions if the error is one of commission, not omission. That is, when we commit an error with purpose, knowing full well it violates established rules, we expect to be punished if caught and we usually accept it. When this occurs, we often accept the value of such discipline and change our behavior.

If the cause of our behavior is ignorance (i.e., omission) and we are punished, we rarely see the value of punishment and will not change our behavior. Indeed, we often seek revenge or take other actions to show our disagreement with the punishment. Whatever the reaction, punishment for errors of omission will likely not cause a change in behavior because behavior is not the cause; the cause is lack of knowledge.

If we disagree with the rules, the cause may be that the rule is inappropriate and needs review or the individual has deviated from accepted thinking for various reasons. It is imperative that we know the causes of what may appear to be inappropriate behavior. If we find that the behavior included a conditional cause of ignorance and an action cause that precipitated the event, then punishment will not afford effective discipline, because ignorance is part of being human. If the cause was to purposely bypass or violate established rules, then punishment may be an effective solution. Even here, make sure you know why the rules were bypassed or ignored.

For example, if bypassing the rules is caused by the long-term failure to enforce or reinforce desired performance, then the responsibility also lies in the leadership and not necessarily in the individual worker. We call this "Common Law," which is not only found in the history of our legal system, it is fundamental to the human condition. If everyone is violating the rules and it is accepted practice, then it is reasonably considered acceptable. In this case, punishment will not be accepted as having value, and behavior will not be modified.

In Common Law situations like this, the performers have been set up to fail by those who are responsible for leading them. A drunk who encourages his child to drink carries a greater burden of responsibly for the consequences of alcoholism than does the child. A leader in any organization must assume the responsibility of setting an example by consistency of purpose or they cannot be called leaders.

Another common cause of inappropriate behavior is the failure to learn. A few people (about 5% of the population) are simply incapable of learning, but a larger number choose not to learn. This cause is evidenced by repeat offenders. If we find an individual does not learn, then reassignment or termination may be the best solution. In this case, however, the solution is not discipline, because it does not seek to change behavior. It seeks to remove the cause of the problem by removing the person who fails to learn. Anyone who thinks they are doing these people a favor by not removing them fails to understand they are actually reinforcing the choice not to learn. Take them out of the cause path.

So yes, punishment is sometimes a viable solution, but it should be applied only when we can be sure that it will prevent recurrence. And that will only occur if we understand the causes. My studies show that we use punishment as a solution about 20% of the time, and it is effective at preventing recurrence less than 1% of the time.

If punishment is used to prevent recurrence, the offending person must know the causes for prevention's sake and must understand the causes in order to accept responsibility. Sometimes we don't see our own willful act to violate established rules. Never rule out the delusionary ability of the human being. It is by far the most powerful attribute we possess. Confront all delusions head on and remember to "fix the cause not the blame."

Responsibility for your own actions requires an understanding of cause and effect relationships. For those who do not understand the cause and effect principle as discussed in this book, accepting responsibility may be a difficult task. I have found one of the greatest secondary effects of training people in these methods is that they come to believe that stuff does not just happen; everything has a cause. With this understanding comes responsibility, accountability, and pride in effective problem solving.

The Root Cause Myth

Common buzz words for problem solving are "root cause analysis," and the term has been around for at least 50 years and in a less formal sense much longer than that. Root causes are the causes that solutions act upon by removing, changing, or controlling them such that the problem does not recur.

With these buzz words, a great myth has been created. When I first became involved in problem solving, I was introduced to all the various methods. I took several training classes, read the few books available, talked to industry experts, and tried to implement the various schemes and tools. When I tried to apply these methods, they didn't help me solve my event-based problems any better than my natural instincts. I looked more closely at the methods and tried to separate the parts that worked from the parts that didn't. Over the years I began to find some things worked much better than others. What I eventually discovered is that the *Root Cause Myth* was causing all the confusion and failed strategies.

The overriding theme of all these methods is the pursuit of a root cause, hence the buzz words: root cause analysis. Funny thing though, there is no accepted definition of a root cause; everyone just makes up their own definition. It took me about seven years of study and teaching root cause analysis to figure out why the definition was so difficult: by focusing on finding the root cause, we presume there is one.

This false premise stems from the following linear thinking: A caused B, and B caused C, and C caused D, on down the alphabet. At some point we arrive at the root cause G and since G effectively caused A we can eliminate the problem if we eliminate G. This common but misguided approach assumes causal relationships are linear and that problems are born from a single source. Perhaps this is some anthropomorphic tendency based on the pattern of life, which appears to have a beginning and end, but as we will learn later, it is fundamentally wrong. This single source of a problem is generally referred to as the root cause and is the basis for most root cause analysis methodologies (The Appendix contains further discussion of other methods). Because these other methods are based on this false premise, they only deliver effective solutions by chance, not by design.

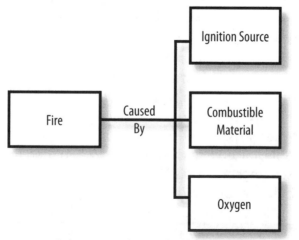

Figure 1.1   Solutions and Causal Relationships

Before I discovered the fallacy of root cause, I taught people how to find the real root cause. After many arguments about whose real root cause was the true root cause, I came to understand that there are many possible root causes; and they are a function of who owns them. More importantly, I began to realize that the focus on root causes is wasted effort because the real goal of problem-solving is to find solutions, not root causes, so we should be first focused on solutions.

To illustrate this, let's look at an example problem: Preventing a fire from occurring. The causal relationship for a fire might look something like Figure 1.1.

Using this simple set of causes, what is the root cause of fire? Think about this for a minute.

Based on your understanding of a root cause, what is the root cause for fire?

If the solution is to remove all ignition sources, then the root cause of the fire is ignition sources. If we decide to control all combustible material in the area, then the root cause is combustible material. If the fire is in a tank like the fuel tank on an airplane and our solution to prevent fire by inerting the tank with non-combustible nitrogen then the root cause is oxygen.

As you can see, in each one of these scenarios, it is the solution that determines the root cause. We must identify one or more solutions that somehow act upon a specific cause to prevent recurrence of the defined problem. It is the solution that defines the root cause, not the other way around as these failed methods would have you think.

When using the conventional wisdom that seeks to first find a root cause we are putting the cart before the horse. By pursuing a root cause, we end up stopping on a single cause that may or may not produce the best solution. The various methods that use this failed strategy are categorical methods which we will discuss later. The thing to remember at this time is that it is not root causes we seek, it is effective solutions.

This does not mean there is no such thing as a root cause, it simply means that root causes cannot be labeled until we decide on which solutions we are going to implement. The root cause is secondary to and contingent upon the solution.

The Illusion of Common Sense and a Single Reality

When the jury in the Oklahoma City bombing trial could not decide on the death penalty for convicted terrorist Terry Nichols, a jurist lamented, "If I learned anything from this, it is that two people can look at the same situation and see two completely different things." Indeed, how could this happen? Where is the common sense? The evidence was obvious, the decision clear. What's wrong with some people anyway? We usually end this line of thinking by concluding that some people just don't have any common sense.

When asked, most of us believe we have our world pretty well figured out and are good problem solvers. We even believe that most of those around us are equally good at problem solving. In fact, we seem to believe that problem solving is the same for everyone. We believe that if we are able to think of it, it must be common to everyone else. Sometimes, when people don't act according to our preconceived ideas, we say they don't have any common sense. We may even question our friendship with them because we certainly don't want to associate with idiots.

Common sense is defined as the common feeling of humanity. With tongue in cheek, it can be defined as that body of knowledge that my friends and I share. In either definition, it is anything but common because we don't have the same friends or the same feelings as the next person. Common sense is often used as an excuse for explaining why others do not "see" things the way we do and then punishing them for it. I once heard a chemical plant manager say, "Since when did our people start checking their common sense at the gate?"

Each one of us is unique, and that uniqueness is caused by our genetic building blocks and the environment in which our perceptions were developed. Exploring why our perception is unique helps us debunk the notion of common sense. Perception exists within each mind and is a four-step process:

1. Receiving data from the senses.
2. Processing the data in the mind to form knowledge.
3. Developing problem-solving strategies.
4. Establishing conclusions and prototypes.

Our Unique Senses

Receiving data from the senses is unique to each one of us. Our sight, hearing, touch, smell, and taste are different than other people—sometimes significantly different. Some people need glasses to see, others don't. Our senses are developed early in life and are a direct function of our environment. Research indicates that children who are visually entertained in the first year of life establish more neural connections and hence have more active minds.[2]

The brain reserves certain areas for each sense. The visual cortex, for example, is located at the rear of the brain, the sensory cortex along the sides, and so forth. As each sense is stimulated, neurological connections are being made in the respective portion of the brain. Patterns are recognized and value assigned to each stimulus in each sensory portion of the brain.

The development of each sensory portion of the brain is a function of the genetic structure of the mind and environmental stimulation. Each sense is on a genetically coded timeline for development. Once that time frame has passed, the sense will all but stop developing.

The acuity of each sense depends on the richness of the environment to which it is exposed during the window of opportunity. For example, if a child is completely blindfolded for the first three to six years of life, the sight portion of the brain will not develop and the child will never see, even though the eyes are completely functional. Physicians have found that covering one eye of an infant for a short period of time (a week or more) will likely cause that eye to be less developed than the other one, resulting in the need for glasses[2] and in a different perspective of the world.

And so, on goes the development of our senses, such that every person senses the world differently and creates his or her own unique sensory perception.

Our Unique Knowledge

As data or information is sensed, it is processed into categories for economy of thought. We assign nouns to things and verbs to actions. Everything is sorted, prioritized, and possibly stored.

When we are young, there is little judgment going on; the mind is like a sponge that simply wants to be stimulated. The more time spent learning, the greater our knowledge. A person who is preoccupied with survival, such as our recent ancestors, had little time left for learning beyond what was required to survive. Today, survival is much easier and knowledge is abundantly available to most people. With this abundance comes a greater diversity of thought.

Over time, sensed data is organized and stored as knowledge. This knowledge is structured and valued in various ways but is always shaped by our environment into a unique perception of the world. For example, in some cultures animal sacrifice is a holy event, in others it is cruelty.

We all have our own interests and abilities based partly on the environment and partly on our genetic makeup. Growing up in Africa with Jane Goodall as your mother would provide you with different knowledge than if you grew up in a poor neighborhood in a large city, such as New York. The resulting personalities and perspectives would also be quite different. While we share many common characteristics, we each possess our own unique knowledge base.

Our Unique Strategies

A key aspect of perception is how we order knowledge. The ordering process is what we call strategies. For example, an infant may learn that crying causes hunger to go away because it causes someone to feed him. From this causal relationship, children may learn the strategy of whining to get their way. Depending on reinforcement from our environment, we will adopt or drop a given strategy.

If we obtain our goals with a given strategy, we will retain it as part of our belief system. Each strategy becomes part of the mind's operating system, and every person uses different strategies for dealing with life's problems. One person may find success in stealing, while another finds failure. Or, in the business world one person may use the strategy of building networks to advance whereas another might use the strategy of working long hours on many projects. Hence, each person will determine the "best" strategy based on their own experiences, where "best" is unique to each person.

Our Unique Conclusions

The mind is continually sensing, ordering, and developing strategies. It is always open to new possibilities but to varying degrees. As adults, we seek validation of existing beliefs (knowledge and strategies) and do not like change. Inherent in our operating system, however, is the prototype strategy. We know from past experience that sometimes things don't happen exactly as they did the time before so we reserve the right to change our belief system. In effect, we naturally establish prototypical truths that are the best we know now but are subject to change given strong enough reasons to do so. For example, for most of us the earth does not move under our feet and this is the truth. Anyone who has experienced an earthquake, however, knows this is not valid—the earth does move and it can move violently. If you have felt the earth move under your feet or have seen a wave in the earth move across a field, your first perception may be one of disbelief, but you soon change your belief system to accommodate the evidence.

We hold our belief systems open to change by the use of a prototypical conclusion. Our unique perception of the world, coupled with our unique interaction strategies, combines to form unique people with unique prototypical truths. All these factors are continuously evolving, some more so than others; but there is clearly no way to be anything but unique individuals. No two people will hold the exact same set of prototypical truths, not even conjoined twins who live in the same environment. Once again, our conclusions cause a unique perception of the world.

Understanding this uniqueness calls into question the notion of common sense. What does it mean to have common sense when not a single person has the same view of the world or holds the same belief system? Indeed, what is real? What is reality? Can we know it? When we use the word reality, we assume that there is a single reality and everyone can see it. By understanding the biological impossibility of perceiving the world the same, the notion of a single reality can now be seen as the illusion it is.

The notion of common sense is therefore an illusion created by the false belief in a single reality. Perhaps this need for a single reality is created by our desire to get along with one another. If we all hold the same beliefs, we could always agree. Whatever the cause, the belief in a single reality is one of the greatest barriers to effective problem solving I have found.

So, if perception is reality and everyone's reality is unique, what is reality or truth? This question of the ages continues to haunt us, but the answer is quite simple if you can grasp the notion of relativity. Everything is relative to our own unique perceptions. We each hold our own truths, and the best we can hope for is to find a way to incorporate others' truths into ours. While we use many tools and strategies for doing this, such as team building, they often fail and understanding is left wanting. Once again, we have been set up to fail.

Groovenation

No, this is not a sixties-era song; it is a human condition of the mind that prevents effective problem solving. Groovenation is a term I created to describe the process of justifying our beliefs. To be groovenated is to hold strong biases and prejudices. It is physiological in origin and is found in our search to validate our existing realities. It is the groove we get in by placing a greater value on familiar things than on differences or change. We have a strong desire to be right in our beliefs, and we continuously seek validation over other possibilities.

As we sense the world, we send all data through what I call a "Delta Checker" from the Greek symbol for difference. The Delta Checker, a learned strategy, checks for differences between what we are sensing and our existing prototypes. If something is the same as previously known, we like it. If it is different, we analyze it and make a value judgment. If we place value on the difference, we continue to scrutinize it. Unfortunately, we have a strong tendency to place high value on anything that mirrors our existing reality and low value on everything else. This tendency is caused by the physiology of the mind. For every thought, idea, sense, or motion, many synapses are fired; and with each firing, the connections become physically stronger both in size and chemical response. Just like building muscles, the more we exercise or stimulate the mind, the stronger it gets.

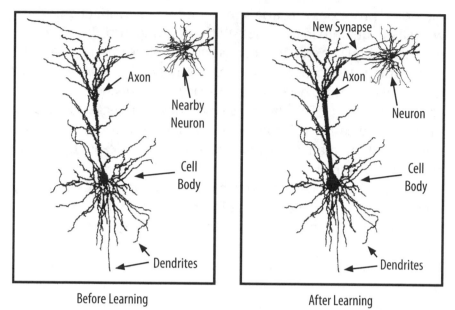

Before Learning   After Learning

Figure 1.2   Impact of Repeated Stimulation on Learning

In the thought-provoking book, Descartes' Error,[3] Antonio R. Damasio, M.D., provides great insights into the working of the mind. Dr. Damasio and others have found the causes of groovenation in the physical nature of the mind. The brain is made up of billions of cells known as neurons, which consist of a cell body, a main output fiber called an axon, and input fibers known as dendrites. These neurons are interconnected in circuits and systems within the brain. Brain functions, including our ideas and thoughts, occur when neurons become active through an electrochemical process. Each time we have a new thought or experience something new, axons and dendrites "connect" via a synapse. If the same thought or experience is repeated, the same physical connections enlarge. Figure 1.2 shows a simplified version of this process.

This is not to suggest that one connection constitutes a specific piece of conscious knowledge. It is much more complex than that, but the observation that these neurological connections occur during learning and actually grow in size with repeated exposure to a given stimulus helps explain part of the physiology behind groovenation. That is, if a larger connection provides a preferential path for a neuron firing, it would explain the physiology of groovenation. Scientists have recently discovered there are other biological processes that also strengthen these connection, which we will not get into in this book. Without going into all the causes, new ideas require new connections and therefore new ideas are at a disadvantage to old ideas. This does not mean we cannot learn, but it does mean we may have to modify existing connections to register contrary thoughts. This happens when new connections become the preferred path from repeated stimulation and connections that don't get stimulated are actually dissolved (physically) by special compounds in the brain[4].

It seems that no matter how hard we try, sometimes it is nearly impossible to pull ourselves out of a groove or rut. This groove can be an idea, a belief, or a habit. Someone who is highly groovenated will remain intransigent even when the path leads directly to a harmful outcome. The kamikaze pilots in World War II or the suicide bombers of modern day radical groups provide a vivid example of a highly groovenated state.

In your daily life, consider the people who judge everything they see and proclaim it right or "WRONG!" Groovenation is a natural state of being too focused on being right while ignoring a broader perspective. It is present in all humans and varies from inconvenience to the paralyzed mind of a fanatic.

Understanding the cause of groovenation can help us understand that it is part of being human. Understanding the physiology behind the process can help us see how easy it is to be brainwashed or to develop an intellect for music, athletics, or whatever we choose. Just like practice makes perfect in sports, repetition of an idea or thought can create a perfect reality that only exists in the mind of the one who created it. It becomes real, regardless of contradictory evidence. Denial is one of our most destructive strategies and groovenation is the cause. (More on Denial in Chapter 6) If we spend our lives trying to validate specific relationships, these relationships will indeed become valid. They become valid because of repeated exposure of the mind to the same conditions. Pick any controversial topic—extraterrestrials, evolution, creationism, or who has the best football team—and you will find proponents that know the "truth" of their position. What they don't understand is that their truth is the result of their own brainwashing.

Final Report

Incident Date: 10/28/04   **Time:** 0817
Report Date: 1/7/05   **Facility:** West

Description of Incident: On October 28, 2004, a contractor electrician was conducting an operational check on elevator ELH-23 at TCH-3-675 when a flashover occurred. The electrician needed to check the door motor and switches on the top of the elevator car, requiring the elevator to remain energized in performing this difficult task. While this was going on, a mail person from Central mail pushed the call button on the first floor ignoring the out of service sign posted over the call buttons. The electrician heard a "buzzer" sound and was able to get clear of the moving parts on the top of the elevator car on which he was working before being injured, but it was a close call. He got control of the car from his location on top of the car, which allowed him to stop the car and exit safely. The main fuse blew and the elevator shut down. Due to multiple parties involved in this incident, extensive discussions and management oversight have occurred. This has caused some delay, however.

Type of Failure: ___Predicted ___Failure
              ___Failure to Secondary Damage __X_ Other

Description of Cause: A critique was held on Nov. 30, 2004, at the incident's location. The FCT electrician involved demonstrated step by step actions taken before the maintenance activity. The Investigation discovered the problem to be human error, and corrective actions are being taken to stop this from happening again.

Corrective Actions:
1. Provide refresher training to all employees on importance of warning signs.
2. Possible testing procedure to include a better lifted leads and jumpers control log.
3. Revise the electrical drawings to show the complete circuit for the elevator controls.
4. Position switches have been ordered to monitor the length of cable.

With the completion of these changes, the problem will not recur.

Root Cause: Defective or Failed Part

Figure 1.3 Example Incident Report

As our brains are conditioned into a physical state by the repeated firing of the synapses, we convince ourselves of the absolute validity of our beliefs. Groovenation presents a formidable challenge to effective problem solving. Because this barrier is deeply ingrained in the human condition, overcoming it is a primary focus of becoming better problem solvers.

Storytelling

Our primary form of communication is through storytelling. Storytelling describes an event that relates people ("who" elements), places ("where" elements), and things ("what" elements) in a linear time frame ("when" elements).

Incident reports provide prime examples of storytelling and its impact on problem solving. Figure 1.3 is an example incident report taken from the manufacturing industry and is typical of 60% to 70% of the many incident reports I have seen, with many being much worse. As you read this example, ask what the problem is, what the causes of the problem are, and if the solutions will prevent recurrence of the stated problem. Remember, this is a typical report. The form is filled out, the boxes are checked, and the categories are defined or discussed.

Is the problem an injury, a near-miss with electricity, something called a flashover, or what? Whatever the problem was, are the causes clearly stated? Do the corrective actions support the statement that this will never happen again? What do "position switches" have to do with anything? Also, notice that the report states that the problem is human error, but then says the root cause is a defective or failed part. Aside from this contradiction, these are cause categories—not causes—to which thousands of solutions could be attached. The proposed "Corrective Actions" discuss training, procedures, incomplete drawings, and a position switch. Since these have nothing to do with human error, additional contradictions are presented. Is there a hidden agenda here? Are these vague references listed on purpose, or is this simply another example of poor problem-solving skills? After the company further investigated this problem, it was shown to be painfully ill defined and the solutions woefully inadequate.

Take note of the nice story in the "Description of Incident." Not only do we write stories in our incident reports, we are told to write stories. Everyone likes a good story. Many companies don't even write the story down; they get together with the decision makers and tell stories to one another, decide which category the problem fits into, and implement their favorite solution.

While entertaining, stories seldom identify causes because they are busy setting the stage of who was where when some action occurred. The basis for any story is a sequence of events starting at some arbitrary point in the past, leading the reader to a significant consequence disguised in a statement like, "The final investigation discovered the root cause to be human behavior." Opinions, or the consensus of a group, are then presented as corrective actions. When corrective actions are disjointed like the ones in this example, the consensus strategy of a committee is probably at work.

Every time I teach a class to a new client, I ask them to send sample event reports. I always find one or more that look like the example discussed here; and when presented to the class, the most common response is that it is typical or not unusual. Supervisors and managers are especially frustrated by such reports but are lost for ways to remedy them. The concern, however, is more than just poor reports; it is poor problem-solving skills that are reinforced by poor report writing and rule-based thinking like filling out a form. Forms subtly tell users to turn off their brains, fill in the blanks, write a good story, check the boxes, and identify the right categories.

Another example of an incident or event report is shown on the next page. As the example shows, the focus is on people, places, and things, occurring as a sequence of events. Few causes are stated. Even the stated root cause discussion is mostly story. We are told that an employee was injured because he fell. He fell because a rubber floor mat slipped, and the root cause was operators leaving the pumps on during breaks, causing oil to leak and leading to a slippery floor and the accident.

An analysis of this report reveals the following:

The root cause, "operators left pumps on," places blame on the operators; and the corrective actions express a pre-established opinion as to what should be done about a condition that somehow may relate to the injury. That is, by making the operators responsible

OSHA Recordable Injury

Discussion: This afternoon an accident that resulted in an OSHA recordable accident at about 7:00 p.m. The employee was working in the yellow room on machine #3. The employee was in a rotating position when the rubber floor mat slipped, causing the employee to fall to his knees, with his left knee hitting the metal legs of the work table. The employee suffered an "abrasion" just below the left knee that was very painful and swollen. I took the employee to the Payton Northeast Hospital emergency room where he was awaiting diagnosis. He contacted his mother to pick him up. I left my name and number and asked him to contact me to let me know how things came out.

I returned to the yellow room to investigate the incident. I asked Diane if anyone was in the area at the time of the incident and she mentioned "Katy, Tom, and Billie." I asked each employee if they had witnessed anything. They indicated they had not. It is apparent that no one actually witnessed the incident. I returned to the area to talk to the lead person, Billie, but she had gone home for the night. My concern is: 1) whether the incident occurred when the employee was alone, 2) is there a policy about employees working alone in the yellow room, and 3) was this the case?

Root Cause: In talking to the lead maintenance person, Ray Longsine, the root cause is "operators leaving the pumps on during breaks." Per Ray, this causes the pumps to operate at all times holding 8.0 psi of pressure which after a prolonged period, causes the oil to get so hot that the foaming is visible through the oil level check window on the pump. Ray said this will only occur when the pumps run for an extended period of time (> 4-5 minutes). He said that all employees have been told not to leave the machines energized while they take breaks. He has even checked machines during breaks and provided feedback to those in violation. Ray also indicated that the problem only surfaced about three months ago when a lot of new training was being done. It appears that this oil problem is not monitored across all shifts and the oil should have been detected long before it caused an accident.

Corrective Action: Have operators be responsible for checking the machines for oil leaks several times during the shifts for leaks. Also, enforce the policy of not leaving the machines on while taking breaks. Ray indicated that John Sisner had looked into a different type of pump, the type that is used on the new machines. Don't know what the status is on that. The pumps are such that they don't go out, they just start leaking oil. Ray indicated he had changed out five pumps in the last two months.

to check for oil leaks, the assumption is that the operators are irresponsible and may even think nothing of working on oil-slickened floors. Once again, the employee has been set up to fail, failed, and then told to be more careful, or in this case, more responsible. All this without ever talking to the injured employee.

If you want to observe a storytelling activity, pick up any newspaper or magazine. Talk to your friends; or the next time you are in a meeting, listen carefully. Or, the next time the President of the United States or any other politician speaks, listen carefully. You will hear all the elements of a story listed above—who, what, where, and when—but you will not hear many causes.

It doesn't matter which culture, country, or education level we observe, we have a common human affliction of poor problem solving, and it is directly proportional to storytelling. The stronger the storytelling culture, the less effective we are at problem solving. A storytelling culture can exist within organizations or within different regions of the country or world.

Storytelling sets us up to fail by ignoring causes and the cause and effect principle, which we will discuss in Chapter 2.

Categorical Thinking

Categorical thinking is caused by the mind's need to order what it perceives. While categorization is a natural process of the mind, the problem comes when we fail to understand how categorization can lead to intellectual laziness. The notion of good and bad is categorical thinking at its most base level. Instead of seeking to understand, we categorize something as good or bad and stop there.

Is it good that the lion eats the gazelle, or is it bad?

Neither, it simply is; and to assign a categorical answer, like bad, we misrepresent the situation by oversimplification. We establish a course of action because if it is bad we are compelled to right the injustice and make it good. If it is good, we can ignore the situation and move on to more bad things.

Please note, it is not the category that causes the problem. Categorization can be a very helpful strategy. The problem categorization creates is the belief that once categorized we can establish certain relationships and then act according to our favorite solutions or stop thinking.

Categorization is strongly linked to storytelling. As the two previous incident reports demonstrate, the causes of the events are in categories. That is, the stated causes represent a group of causes, not a specific cause we can act upon. Here are some examples I hear often: "The cause of 95% of all industry accidents is human behavior." "Corporations have caused most of the environmental disasters in the world." "The pump failed because it was worn out." In these three examples, human behavior, corporations, and worn-out equipment are all categorical causes. Stopping at categories like "worn out" usually leads to ineffective solutions such as replacing the equipment. Solutions based on categorical causes fail to correct more fundamental causes like the cause of wear. The end result is recurrence of the event.

The Garbage Solution is my favorite example of categorical thinking. As we go through the day, we have to assess the value of many objects. Some objects have great value; others, like a banana peel, may have none. Our solution: put these no-value things in the garbage and someone will make them disappear. With this strategy we solve many problems with one solution. The Garbage Solution has us put many problems into one category and solve them with one solution. It appears to be effective and is used extensively in all aspects of our lives. In business we seek "the biggest bang for the buck!" We put as many problems as we can into a category and then solve it with one solution.

The danger of this strategy is that it doesn't address each individual problem and may cause many other problems. Examples of this categorical thinking are all around us. Lumber mills used to burn all their scrap until they finally realized they were polluting the air and wasting valuable raw materials. The solution to pollution used to be dilution until we looked closer at its effects. Now most pollutants, including our garbage, are evaluated for cause and effect relationships and individually controlled.

The following is another common example of categorical thinking. We seek to know where people come from, their education level, religion, or political alignment, in order to know "who" they are. Is there anyone among us who doesn't do this? For example, if you are not well educated and I am, then I can draw certain conclusions and act in a certain way. Such an inference precludes me from knowing the real you, yet we use this strategy daily because we don't have time to do otherwise.

Categorical thinking creates another significant problem for data collection and analysis. When interacting with others, we assume there is a single reality and therefore our categories are identical to theirs. They are not. I have run hundreds of experiments where I ask students to categorize a list of causes. When completed, we compare notes and find the same causes in completely different categories. The magnitude of this discovery is significant. Every database that has ever been created from the input of more than one person, like most accident reports, has questionable data. I am opposed to categorization, as will be discussed later, but if it must be done, I recommend all categorization be performed by one person or a small cadre of like-minded people. In any database, consistently comparing apples with apples and oranges with oranges is essential. We are not doing this now because we believe in the notion of a single reality and that everyone "sees" the same world.

Categorization is part of our natural operating system. In the process of categorizing, we assign value that establishes our biases and prejudices. By not recognizing the danger these prejudices bring, we set ourselves up to fail with ineffective problem-solving strategies.

Causes of Ineffective Problem Solving

As we review the many examples of ineffective problem solving, we discover a recurring trend—effective solutions are not found because of three things:

1. Incomplete problem definition.
2. Unknown causal relationships.
3. A focus on solutions.

Let's assess each one of these factors.

Incomplete Problem Definition

Incomplete problem definition is caused by the false belief that the problem is obvious and the subsequent rush to find a solution. The belief that the problem is obvious is caused by the belief in a single reality discussed above and the notion that we all think the same (common sense).

Observing thousands of decision-making meetings, I found that most meetings start with a statement something like, "Thank you for coming. I think everyone pretty much knows what the problem is, so does anyone have any ideas how we can prevent this from happening again?" The leader looks around the room and sees everyone nodding their heads in approval, so continues.

Everyone is nodding in approval but is thinking many different things. The operations manager knows the problem was caused by poor maintenance. The maintenance manager knows it was those darn operators again, and the facility manager knows somebody screwed up again. A few people will offer their solutions and the arguing commences. The battle for the right answer begins and the person with the best story—usually the boss or whoever is considered an "expert"—wins. Quite often the discussion ends with the boss or expert expressing their reality—everyone agrees, and they move to implement that person's wishes.

Most people have been in this meeting. As you can see, the problem is not defined, causes are ignored, and the focus is on sharing our favorite solutions to show everyone else how smart we are. Little or no synergy occurs; the problem is never fully understood and it happens again. When it does recur, we reconvene the meeting, implement another favorite solution, and the cycle continues.

We track and trend how many times we have certain types of failures and create beautiful graphs and charts showing the various "root causes," but the problems keep recurring. We become so engrossed in tracking and trending causes that we do not realize we are failing miserably. Effective problem solving means the problem never happens again.

Unknown Causal Relationships

Causal relationships often remain unknown because we do not seem to think causally. Instead, we communicate by telling each other stories, and the inferences surrounding the stories pass as causes. We infer that hardware wore out by recommending replacement. We talk in terms of human error, lack of training, and other categorical causes like management being less than adequate. Instead of continuing to ask "why" to our point of ignorance, we stop at causes that align with a favorite solution. Why did the hardware wear out? This, along with several more "whys," is required to adequately understand the problem. Our ignorance of the cause and effect principle as discussed in the next chapter is perhaps the largest contribution to incomplete causal analysis.

A Focus on Solutions

By focusing on solutions without clearly defining the problem and its causes, we often find ourselves solving the wrong problem. Today, most people still believe the Exxon Valdez oil spill was caused by a drunk captain. The federal government focused on this issue as the problem and proceeded to penalize the Exxon Corporation as the main solution. Monetary penalties did nothing that I am aware of to prevent future oil spills. (Note: While this is a sad truth, the purpose of the legal system is to punish, not prevent recurrence.) Industry efforts to utilize double-hulled tankers has done more to prevent recurrence than the fines levied on Exxon.

Focusing on solutions is caused by many things, but chief among them is groovenation. As discussed earlier, our brains are wired such that we search for what we already know and when we find it, we validate the rightness of our search and cease to look any further. My daddy drove a Ford and his daddy drove a Ford; therefore, I will always drive a Ford. This kind of logic is a natural mental process whereby we seek the familiar and call it "right" or "real." We tell our children "wait until you grow up and have to face the 'real world.'" The "real" world will be their world, not ours, and it will be much different. This tendency to seek the familiar is called the favorite solution mindset, and it prevents effective problem solving most of the time.

Authoritative and goal-driven personalities contribute to this drive to find a quick solution. The "Ready, Shoot, Aim!" personalities are hard to deal with because they are ensconced in categorical thinking and buoyed by past successes (groovenation). It is important to remember that groovenation is a strong physiological force and that even the best and the brightest can succumb to it.

Set Up to Fail

By sharing my observations of the current state of problem solving, I have attempted to show that we are set up to fail by the processes and strategies we use. Groovenation and the unintended pursuit of the ignorance it creates are the driving factors in this setup. When coupled with the misguided belief in a single reality and the illusion of common sense that this creates, these strategies set the formula for ineffective problem solving.

What we need is some way to express every stakeholder's perspective in a way that complements the learning process. The diversity each person brings to the table provides the opportunity to see a bigger and clearer picture of each situation. Unfortunately, we often look on individuality as counterproductive to effective teamwork, when in fact it is our greatest strength. Conformity, not individuality, is the enemy of effective problem solving. When we conform, we align our thoughts into one point of view.

With many diverse thoughts, there is no limit to what we can accomplish. This principle can be illustrated by the bed of nails supporting the East Indian guru. The nails together as individual units support the body. If they were all aligned in a row, they would be a big skewer. The guru is not skewered because his weight is distributed over many individual points, each providing a small support and together supporting all the weight. Together, as individuals, we can create a common view greater than any individual reality.

Up until now, we have never had the tools to allow us to create a common view or understanding—to form what I call a common reality. Creating this common reality is what Apollo Root Cause Analysis is all about. By creating a common reality made up of all perspectives, we are able to break out of the illusion of common sense and thus prevent the usual arguing that prevails. It also helps us break the bonds of groovenation by presenting legitimate realities heretofore unknown.

Given proper tools that overcome the handicaps of existing strategies like common sense, single reality, storytelling, categorization, and the pursuit of a root cause, we can find effective solutions to everyday problems almost every time. These tools are based on the cause and effect principle discussed in the next chapter, and they will help us break the bonds of ignorance.

References

1. Performed by James M. Stoutenburg as part of his graduate work in Instructional and Performance Technology at Boise State University, Boise, Idaho, 1994.
2. The Amazing Brain, Robert Ornstein and Richard F. Thompson, Houghton Mifflin Co., Boston, 1984.
3. Descartes' Error, Antonio R. Damasio, M.D., Grosset & Putnam, New York, 1994.
4. Receptors, Richard M. Restak, M.D., Bantam Books, New York, 1994.

2

Understanding the Cause and Effect Principle

Nothing happens without a cause. The notion of cause and effect is fundamental to all philosophies and major religions and still we hold to the whimsical adage that "stuff just happens." It seems we are incapable of admitting we don't know. There are no such things as mystics and magic, there are only cause and effect and the unknown. As John F. Kennedy said, "Things don't just happen, they are made to happen." Cause and effect relationships govern everything that happens and as such are the path to effective problem solving. By knowing the causes, we can find some that are within our control and then change or modify them to meet our goals and objectives. For at least 4,500 years, mankind has used the notion of causation to express human events (Van Doren, 1991). Unfortunately, we have failed to differentiate the immense power of the cause and effect principle from the simple notion of causation. From Socrates and Aristotle to St. Thomas Aquinas to Carl Jung and more recently Peter M. Senge, many minds have pondered the idea of causation and built upon prior knowledge yet never cracked the code that is revealed herein. This chapter will take you on a journey into the depths of causation like never before documented. As we pull back the veil, we see four important characteristics of the cause and effect principle:

- *Cause and effect are the same thing.*
- *Causes and effects are part of an infinite continuum of causes.*

- *Each effect has at least two causes in the form of actions and conditions.*
- *An effect exists only if its causes exist at the same point in time and space.*

We will examine each of these characteristics so that we can build a set of tools that uses them to understand and document reality in a totally new way of thinking.

While teaching a class in a small town in Georgia, I was eating dinner one evening at a local restaurant. Sitting alone, I was busy watching people. A young family and their friends were seated at the table next to me. They had a small, perhaps nine-month-old, daughter seated in a highchair near her father. As the adults talked, the child was experimenting with a spoon. She banged it on the top of her highchair, licked it, and banged some more. In time, she leaned over the side of her chair and holding the spoon at arm's length, let it go. As it fell to the floor and bounced, she was immediately amazed. She looked around at the adults to see if they had seen this incredible event. They, of course, had missed it. In fact, she noticed they were paying no attention to her incredible discoveries. "What was the matter with them?" I read on her face.

With an outstretched arm and a grimace on her face, she reached for the spoon to no avail. After a few grunts and wanting cries, her father noticed her and returned the spoon to her table top. She smiled and returned to her play. After a few bangs, she decided to try the spoon drop experiment again. Again, it dropped straight down. It did not float upward like those big round colored objects she sometimes plays with; this thing went straight down and bounced on the floor. Again, her face said it all. "This is really cool! Did you guys see that?" Looking up for acknowledgment, she seemed amazed at their total disregard for the profundity of her experiments. Again, she motioned and cried for the return of her object so she could further test the limits of her understanding. As the evening continued, she pestered her parents for the fallen spoon and proved that solid objects when released at height will always fall to the floor—it didn't matter if it was a spoon or mashed potatoes, stuff always went in the same direction.

As I watched this simple event, I saw a child learn about the law of gravity. But there was much more going on here. She was practicing a more fundamental life strategy. She was using her ability to control things and people to advance her understanding of the world.

And, isn't this what we all do? We control things, and we control people to accomplish our goals. In a moment of clarity, I realized that controlling causes is one of our most basic operating strategies. In the process of learning, we identify causal relationships, such as things always fall down; and by controlling certain causes, we are able to accomplish our goals. We learn that to obtain a desired effect we can act upon an object or person, and the effect will be caused to happen. Like the little girl, we may learn that if we whine enough, somebody will fill our need. The more specific knowledge we have about cause and effect relationships, coupled with our ability to act upon the causes within the relationships, the better our problem-solving skills. No matter how complex the causal relationships, be they mere feelings or hard scientific facts, the problem-solving process is always the same.

In the past, scholars tried to understand causation by labeling and categorizing different kinds of causes. Attorneys use proximate cause and probable cause. Safety engineers use surface causes, causal factors and root causes. Aristotle had his four causes—efficient, material, formal, and final, which make no sense at all in today's world. By categorizing we create boundaries or boxes that define the category based on our own belief system. Because we all have different belief systems, as we learned in Chapter One, categorization models immediately set up a quarrelsome environment. To avoid this, it is my goal here is to discuss the characteristics of causes and effects without categorizing different types.

So, what is a cause and what is an effect, but more importantly, what is their relationship to reality? This simple notion of cause and effect is easy enough to grasp as the child did in the spoon drop experiment. However, as we will discover in this chapter, there is much more to this fundamental idea than has ever been explained. Let's look at the four characteristics of the cause and effect principle.

Characteristics of Cause and Effect

The cause and effect principle has four important characteristics:
1. Cause and effect are the same thing.
2. Causes and effects are part of an infinite continuum of causes.
3. Each effect has at least two causes in the form of actions and conditions.
4. An effect exists only if its causes exist at the same point in time and space.

Cause and Effect Are the Same Thing

When we look closely at causes and effects, we see that a "cause" and an "effect" are the same thing. They differ only by how we perceive them in time. When we start with an effect and ask why it occurred, we find a cause; but if we ask "why" again, what was just now a cause becomes an effect. This is shown in Figure 2.1

Effects				Causes
Injury	⇨	Caused By	⇨	Fall
Fall	⇨	Caused By	⇨	Slipped
Slipped	⇨	Caused By	⇨	Wet Surface
Wet Surface	⇨	Caused By	⇨	Leaky Valve
Leaky Valve	⇨	Caused By	⇨	Seal Failure
Seal Failure	⇨	Caused By	⇨	Not Maintained

Figure 2.1. Injury Example

by listing a column of effects and a column of causes. (Read left to right, top to bottom.)

Notice how the cause of one thing becomes the effect when you ask "why" again. The cause of the "Injury" was a "Fall," and when you ask why "Fall," it changes to an effect and the cause is "Slipped." This relationship continues as long as we continue to ask why.

When asking why of any given effect, we may not always agree on the answer because as we learned in Chapter One, everything is relative to our own perspectives. Others may perceive a cause or effect differently or more deeply if they have a greater understanding of the causal relationships. For example, we know we have a cold when we ache and cough, whereas a doctor knows we have a cold when he or she can observe a virus on a microscope slide. The effect is the same, but the knowledge of the causes is significantly different depending on perception and knowledge.

Knowing that cause and effect are the same thing only viewed from a different perspective in time helps us understand one reason why people can look at the same situation and see different problems. They are actually perceiving different time segments of the same event. If we treat each perspective as a different piece of a jigsaw puzzle, we can stop the usual arguing and work on putting the different pieces together.

By understanding that a cause and effect are the same thing only from a different perspective, we get a glimpse of the next characteristic.

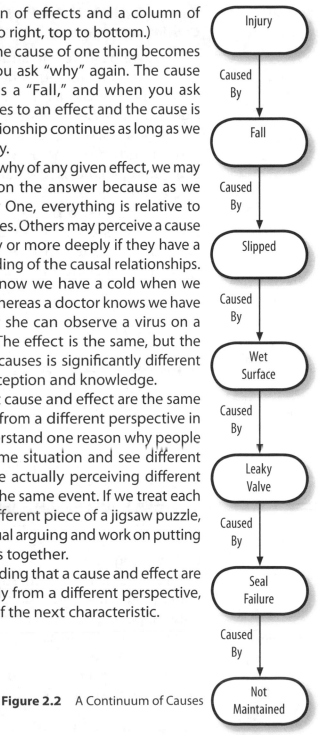

Figure 2.2 A Continuum of Causes

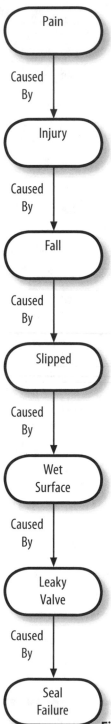

Causes and Effects Are Part of an Infinite Continuum of Causes

As we observe the structure of the cause chain created by asking "why," we are drawn to a linear path of causes. The causes presented in Figure 2.1 have been rearranged in Figure 2.2 to represent a chain of causes. This chain of causes seems to go on as long as we keep asking "why" and getting answers, so where does it start and where does it end?

In event-based problem solving, we always start with an effect of consequence that we want to keep from recurring and end at our point of ignorance. Our point of ignorance is where we can honestly admit we don't know "why."

Presented with a reality that has a never-ending set of causes is something we have great difficulty accepting and probably explains why we stop asking why at an early age and pursue simpler strategies like categorization, and storytelling (more on this later). Designed to find the right answer, the human mind simply cannot deal with not knowing so we create answers when there are none[2]. This is particularly true in group-settings because we don't want to look stupid[3].

Where we begin is a function of our own perspective. If we are the person responsible for valve maintenance in this example, we may choose to start asking "why" with the leaky valve or possibly the seal failure. If we are the safety engineer, our primary interest would be in preventing the injury from happening again, so we would probably start with injury and begin asking "why."

What if we were the injured person? Our interest may be the pain, so our focus would start before "injury" at the effect of "pain"; and we would

Figure 2.3   New Continuum of Causes

have a chain of causes that starts with "pain," as shown in Figure 2.3. For our convenience, we are going to call this starting point the "primary effect."

> A primary effect is any effect that we want to prevent from occurring.

The primary effect is not a universal point that we must somehow discover. It is a point in the cause chain where we choose to focus and begin asking "why." This point can be changed anytime we need to change our focus. We may have more than one primary effect for a given event, which will be discussed later.

Knowing that causes and effects are part of an infinite continuum of causes helps us understand that no matter where we start our problem analysis, we are always in the middle of a chain of causes. This helps us understand that there is no right place to start. Like a jigsaw puzzle, we can start the problem solving process anywhere and still end up with a complete picture. This avoids the usual arguments over who is right and allows us to focus on finding causes. Instead of arguing over what the problem is, like we normally do, we can know that all causes are connected somehow in time and we just need to figure out the relationships.

Each Effect Has at Least Two Causes in the Form of Actions and Conditions

Closely tied with the characteristic of the infinite continuum of causes is the characteristic that causes are not part of a linear chain as depicted earlier, but more like a fishnet. As Figure 2.4 shows, we begin to see that each effect has two or more causes and the

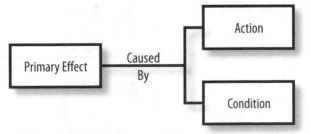

Figure 2.4 The Elemental Causal Set

causes come in the form of conditions and actions. That is, some causes (conditions) are in a passive state, like the air we breathe, while other causes (actions) seem to be in motion or otherwise active. The fundamental element of all that happens is a single causal relationship made up of an effect that is caused by at least one conditional cause, and one action cause.

Definitions:

Actions are momentary causes that bring conditions together to cause an effect, sometimes called action causes.

Conditions are causes that exist over time prior to the companion action.

Elemental Causal Set is the fundamental causal element of all that happens. It is made up of an effect and its immediate causes that represent a single causal relationship. The causes consist of an action and one or more conditions. Causal sets, like causes, cannot exist alone. They are part of a continuum of causes with no beginning or end.

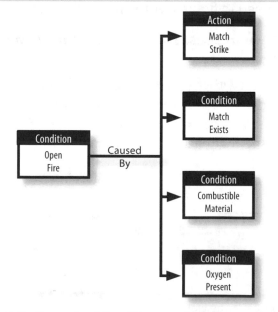

Figure 2.5   Example of Conditions and Actions

By understanding this characteristic, we can look for two or more causes each time we ask "why." Actions are the causes we most easily recognize, while conditions are causes often ignored or beyond our knowledge. If we are able to see the conditions, we often find that several conditions come together with an action to cause some effect, as in the fire example of Figure 2.5.

As we ask "why Open Fire?" we see that three conditions exist in the form of Match Exists, Combustible Material, and Oxygen. The fire is created at the moment the Match Strike occurs. In this example, a match strike is the action; and as soon as that match

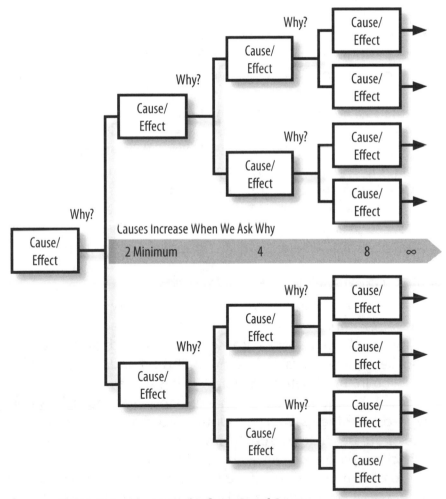

Figure 2.6. Theoretical Infinite Set of Causes

strikes, we have a fire. It takes all three conditions and one action to create the fire.

As we begin to explore the possibilities of this relationship coupled with the second characteristic, we begin to see that causes are part of an ever-expanding infinite set. As Figure 2.6 shows, each time we ask "why," we get two or more causes, resulting in an ever-expanding set of causes. If a fire has four causes and if each of those causes has four causes, then the total set of causes grows exponentially to infinite proportions.

As we look at this ever-expanding set of causes, we are immediately overwhelmed with too much information and the mind can't handle it.

The reason we don't see the infinite set of causes in the real world is because we have learned to filter out most of the causes. We do this quite naturally by assigning priorities and focusing on certain cause paths. We discriminate by allowing only certain causes to be recognized in our conscious mind. The infinite set is there nonetheless. In fact, you could say that it defines reality, but we only see parts of it because we are limited by our knowledge, lack of interest, time available, favorite solutions, the natural limitations of our mind and our language does not allow us to see it. All these filters stifle the questioning attitude we were born with.

If we examine each of these limitations, we see that our level of knowledge limits our ability to know many causes. For example, if we were to ask "why gravity," our ignorance prevents an answer and therefore we cannot continue down this cause path. We must stop and say "I don't know."

Our level of interest also determines our ability to know causes. In the fire example, where oxygen was listed as a cause, we may ask "why oxygen," but we don't because we are met with the immediate response of "who cares?" We know that this condition cannot be controlled in this situation and hence it has no value to us.

Lack of time keeps us from exploring every causal path in day-to-day problems. We limit our time according to our sense of value or our desire to pursue the problem. This leads to a strategy of checking past experiences to see if we have encountered the same problem in the past. If we have, we tend to search for the solutions that worked before and implement them. Quite often, we do not clearly identify the problem or spend time understanding the causes. We simply identify the problem categorically, such as human error, and impose our favorite solution, such as punishment or retraining.

Physical limitations of the mind limit our ability to hold very many thoughts or ideas at the same time. George Miller, in a 1956 article in Psychological Review, first showed that adults can only hold about seven pieces of information in the conscious mind at the same time. For example, we can usually add a few numbers together in our minds without resorting to pencil and paper: 46 + 54 = 100. Likewise, it is fairly easy to remember a seven-digit phone number, but a ten-digit long-distance number or adding several three-digit numbers usually brings out the pencil. Our conscious mental capacity is limited to a small number of thoughts, and yet we attempt to solve incredibly complex issues without writing them down. In the process we forget details and key pieces of information.

This ability to handle only seven ideas may explain why some people believe the root cause appears after asking seven "whys." Although we have incredible storage capacity, our working memory and current conscious thinking are very limited.

With this severe limitation and the problems language presents (more on this in minute), we need some kind of tool or language aid to help keep our thoughts in front of us. This tool would have to allow an infinite set of ideas to be represented. It would have to be clear and simple to use. If we could develop this tool for problem solving and somehow identify all the causes of an event, we could use it to help decide how best to solve our more complex problems. Prior to Apollo Root Cause Analysis this tool has never existed. By first understanding the cause and effect principle, we can devise a set of tools that will allow us to overcome this handicap.

It is important to remember that while our minds naturally filter out or never know many of the causes of a problem, the causes are there nonetheless. Perhaps the single biggest lesson I have learned from all my studies of human problem solving is that we must be humble above all things. When faced with the infinite set, the only thing I am sure of is that we don't have the slightest idea what is really going on.

An Effect Exists Only If Its Causes Exist at the Same Point in Time and Space

Cause and effect relationships exist with or without the human mind, but we perceive them relative to time and space. From observation we see that an effect exists only if its causes exist at the same point in time and space. For example, the little girl's spoon fell because of at least three causes: gravity, the condition of holding the spoon at some height, and her action of letting it go. If these causes did not exist at the same time and space, the spoon would not fall. If the spoon is on the floor, it is in a different space and cannot fall; or if the girl never let go, the spoon never would have fallen.

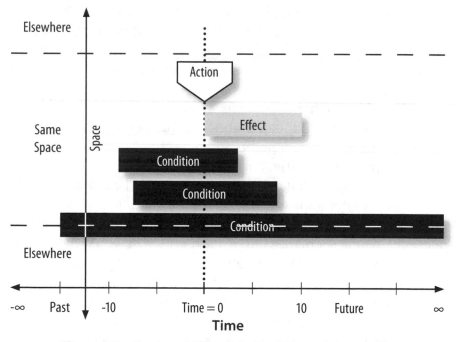

Figure 2.7. Cause and Effect Relationships vs. Space and Time

Every effect we observe in the physical world is caused by momentary action causes coming together with existing conditional causes in the same relative space. A causal relationship is made up of conditional causes with a history of existence over time combining with another cause at some instant in time to create an effect. If we were able to see the world in stop action, we could see, for example, a nail held in place by a hand and a hammer's head striking the nail to cause the effect of two boards nailed together. The nail, hand, hammer, carpenter, strong arm, and wood all exist as conditional causes at the same relative place at the instant the swinging hammer head strikes the nail.

Figure 2.7 shows a generic representation of how several conditional causes come into being in time and are located in the same space. At the instant an action occurs, an effect is created. Change the space or the time, and the effect will not be created.

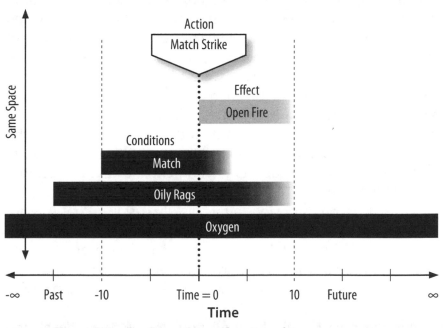

Figure 2.8. Fire Example as a function of time

When asking "why" of a primary effect, our linear thinking usually only provides one answer. However, as discussed, every effect is created by at least two causes (conditional and action) coming together. These conditional and action causes must each have the same "when" and "where" associated with them for the effect to occur. If we do not see this time/space relationship, the causal relationship is not valid.

One of the greatest difficulties in understanding this time-space relationship is the fact that we do not see our world in stop action. The world we perceive is one continuous linear set of causes, all acting together like the frames of a motion picture. Our language even prevents us from expressing our thoughts in anything other than a linear time-based sequence. For example, inside the raging fire are many unseen causes coming and going at a rapid pace. If we step back and look at the big picture, we might see something different, as the following example demonstrates.

Since oxygen has existed on this planet for about 2.3 billion years and is always present in open space, we show it in Figure 2.8 as extending over a long period of time. The oily rags probably have only existed for a short period of time. Assuming the matches existed in the room near the rags for some shorter period of time, one of the matches is struck next to the rags; and we have the effect of an open fire. We could say that actions are causes that bring conditions together, as long as we understand that "bring together" does not always imply physical movement. Conditions are causes that exist prior to an action and are required for the effect to occur. Like the ingredients of soup, each component is a condition; and it isn't soup until the cook says it's soup. At that moment, the mixture becomes soup until it is eaten.

What becomes obvious after this discussion is that it is not easy to communicate these simple concepts because everything is relative; and our minds have difficulty processing more than one relative concept at a time, which in turn is reflected in our language. All modern languages propel us along a linear time line from past to present. They do not allow for branches of conditions and actions. I suspect that if we thought this way, language would have developed to allow discussion of the infinite set of causes, but we are really very primitive creatures and like to keep things simple. Even the notion of infinity is difficult for most people, so it is understandable why our language reflects a simple linear causal-thinking pattern without conditional and action branches.

We must get beyond our linear thinking to understand that actions are momentary causes and there is no effect without them. If we are to overcome this built-in barrier, we need a simple tool that allows us to share branched thinking with others. We will learn about this tool in the next chapter.

Inside the Cause and Effect Principle: Baby Steps

If we look inside the dynamics of causal relationships, we begin to see that not only does the infinite continuum of causes expand

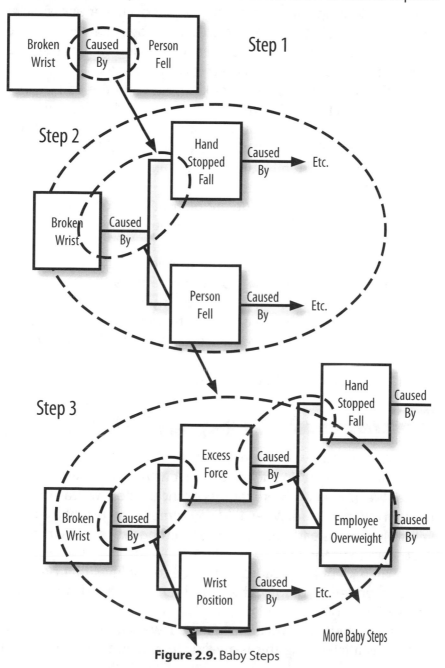

Figure 2.9. Baby Steps

along both ends of the time line and exponentially each time we ask "why," but that there are always causes between the causes. The limitations of our ability to understand the infinite set also apply to our inability to know all the causes between the causes. Every causal relationship can be broken down into smaller and smaller steps. I call these causes "baby-step causes" because they are like the baby steps we take in the process of learning to run. They are there but are forgotten or unknown to all observers.

Baby steps are found by looking between the causes, but they are often illusive. The more we ask "why," the closer we get to understanding specific causal relationships; but the fact is, we will never know all of them. A good example of this is the use of antibiotics today. When first introduced, antibiotics could kill just about any bacterium we wanted eradicated. Today, some bacterium can only be controlled by one antibiotic, and scientists estimate that soon this type of bacterium will no longer be destroyed by any antibiotic.

When scientists first began using antibiotics, they developed a theoretical model of how a bacterium affects the body. After understanding most of the causal relationships, they found a way to control some of the causes so that the bacterium is killed inside the body. This produced the desired effect of returning to a healthy state, but other causes were acting as well. The bacteria are continually evolving and changing their genetic makeup as a result of environmental influences. Scientists continue to redefine their theoretical model; but the fact remains, they do not understand all the causal relationships.

We often know enough about the causal relationships of a system to cause certain effects to exist, but we do not know all the causes. We must know more causes of cancer before we can hope to have an effective cure. When that day comes, our current methods will likely seem as barbaric as the bloodletting procedures from the past seem to us today.

When identifying causes, try to go to the level that provides the best understanding of the causal relationship. This can vary, depending upon our needs.

In our injury example of Figure 2.2, we said the cause of the injury was a fall. While this is a valid statement, there are several possible causes between injury and fall. The example in Figure 2.9 shows the causes between causes and the branches in the cause path. The closer we look and the more we ask "why," the more we find causes between the causes. We begin to understand that not only do causes exist in a continuum that can expand exponentially and have any number of branches, but that there are always causes between the causes. Realizing this begs the question: How far should we go when asking "why"? Always go to your point of ignorance or until you decide to stop. The decision to stop should be based on the significance of the problem and your ability to find the best solution. The concept of defining the significance of a problem is discussed in Chapter 3. If you don't find a fantastic solution, then come back to your cause chart and look for more causes. Identifying solutions will be discussed in detail in Chapter 4.

Linear Language, Linear Thinking

With a new appreciation for cause and effect relationships, let's take a deeper look at storytelling and language. Stories, our primary form of communications, conflict with the cause and effect principle in three ways:

1. Stories start in the past while causal relationships start with the present.
2. Stories are linear while causal relationships follow the branches of the infinite set.
3. Stories use inference to communicate meaning, while problems are known by sensed causal relationships.

Let's examine a simple little story to see how detrimental these conflicts are.

The little handicapped boy lost control of the run-down wagon and it took off down the hill on a wild ride until it hit the little blind girl next to the drinking fountain by Mrs. Goodwin. The little boy was in the wagon the whole way but was not injured. The boy's mother should never have left him unsupervised. The root cause of the girl's injury was human error.

Stories Start in the Past

As you can see, the story starts in the past at the top of the hill and progresses through time from the past to the present, from the beginning of the ride to the end, from the safe condition to the stated problem of the injury. The conflict this creates is that by going from past to present we do not see the branched causal relationships of actions and conditions. If we could know every cause of this injury example, we would see a diagram of cause and effect relationships similar to Figure 2.6. That is, we would see a set of ever-expanding causes starting with the injury and proceeding into the past. To express what we know causally in story format, we would first need to express all the causes on the right side of the diagram, i.e., starting from the past. Our language and the rules of storytelling simply do not allow for this. We cannot express 16 causes and then tell what they caused and so on. No one would sit still for a story told this way because stories are about people, places, and things as a linear function of time.

Stories Are Linear

As we look at this simple story, or any story, we find our language restricts us to a linear path through time and space. Stories go from A to B to C, linearly through time without regard for the order of causal relationships. We are told of the little boy losing control of the wagon as it goes down the hill and strikes the little blind girl. There are no ever-expanding set of branched causes expressed like those in Figure 2.6.

We have the ability to escape this linearity and express branches if we use the words "and," and "or," but the rules of grammar tell us not to use these connecting words excessively. The best we can accomplish is one or two branches for each sentence. The conflict arises because the cause and effect principle dictates an infinite set of causes for everything that happens, while stories are created and expressed linearly.

Storytelling, whether it is ancient history or a recent event description is a linear understanding of an event in a time sequence from past to present, and totally ignores the cause and effect principle. Because we do not understand the branched causes of the infinite set, we use our own understanding of cause, which is generally to follow the action causes. (See Manuele[4], Ch 8, 1997, for an extensive discussion of this effect.) Because we typically fail to see conditions as causes, we ignore them and primarily focus on a linear set of action causes, which are often initiated by people.

Stories Use Inference to Communicate Causes

Since good stories seem to provide us with a valid perception of what happened, we need to question how this can occur in light of the above conflicts. The key word here is "perception." When we read or hear a story, our mind provides most of the information (Carter[2], 1999; p 149). As we read the words, we are busy creating images in our mind's eye. These images are created from past experience and assembled into a sequence of events.

Because the sequence of events (the story) does not express the branched causes of the infinite set, we must make up for it somehow and we do this by inference. We infer causes within the story that are not stated. For example, we read that the little handicapped boy lost control of the wagon. Since no cause is stated for how he lost control, we can infer anything our mind will provide, and we do just that if questioned about it.

Furthermore, stories infer cause by the use of prepositions like "in," "on," "with," etc. Prepositions and conjunctions by definition infer a relationship between words, and the relationship is left to the reader. The word "and" is often used to mean "caused." In this story we read that the boy lost control of the wagon *and* it took off down the hill, meaning the loss of control caused the wagon to take off down the hill. Within this "and" is the potential for many causal relationships and they are left for the reader to interpret. For me, the "and" between *lost control* and *took off down the hill* is obviously a broken steering mechanism, while someone else may picture a paraplegic little boy as the cause, the next person sees the wagon wheel strike a rock that causes the wagon to veer sharply. Because we do not express what is happening causally, each word in the story provides the reader with the opportunity to know more about the event than is stated—to interpret the situation from their own biased mind, which is not necessarily what actually happened.

In the end, each one of us thinks we know what happened but we really don't because stories do not express the full set of causal relationships. Our linear language and the linear thinking behind it prevents us from knowing and expressing what really happens in any given situation. Couple this with the notion of common sense and the false belief in a single reality and you have the causes for miscommunication and ineffective problem solving that is so prevalent in every human endeavor. You have the cause of why almost every decision-making meeting includes conflict and arguments.

What we need is some way to communicate and assemble the causal relationships that each one of us brings to the table. By breaking away from storytelling and knowing the causal set for the problem at hand, we can find effective solutions every time.

Understanding the Cause and Effect Principle

By understanding the cause and effect principle, we can represent any situation using causes. By knowing that causes are part of an infinite continuum, we know that no matter where we start working on a problem we are always in the middle. Since cause and effect are the same thing, we can move forward or backward along the cause continuum as we learn more about the causal relationships of our problem. With this flexibility we eliminate the typical bickering about what the problem really is. All ideas are accepted and aligned causally in time from present to past.

By looking for an action cause and conditional causes of each effect, we gain a much better picture of the problem and its causes. By understanding the notion of an infinite set of causes, we are no longer restricted by our own paradigms. We know that each cause is like a piece of a puzzle and each person's perspective provides insight into the causes. With this understanding, the task becomes one of assembling all known causes rather than bickering to determine who has the correct ones.

By understanding the four characteristics of the cause and effect principle, we can create tools that will help us break out of the old paradigms of linear and categorical thinking. This new tool set will allow us to escape the death grip of a single reality and encourage everyone involved to share their ideas and thoughts. In doing so, everyone will come together with their own realities to form the common reality we need. Being principle based, this tool set will work on any problem. This tool set is presented in Chapter 3.

References

1. Van Doren, Charles, 1991, *A History of Knowledge*, Ballantine Books, New York, NY

2. Carter, Rita, 1999, *Mapping The Mind*, University of California Press, Los Angeles, London

3. Goleman, Daniel, 1995, *Emotional Intelligence*, Bantam Books, New York, NY

4. Manuele, Fred A, 1997, *On The Practice Of Safety*, Second Edition, John Wiley & Sons, New York, NY

3
Effective Problem Solving Defined

An idea is more powerful than an army, but neither can succeed without tools.

Effective problem solving is built on the cause and effect principle, but we need some simple tools to help us implement this principle. The problem-solving process discussed in this book is a synergistic event that transforms ordinary problem solving into an adventure in learning by all stakeholders. The tool set you are about to learn in the rest of this book is based on the following four phases:

1. Define the Problem.
2. Create a Realitychart[1] using RealityCharting® software.
3. Identify Effective Solutions.
4. Implement the Best Solutions.

This chapter focuses on the first two phases. Phases 3 and 4 are covered in Chapters 4 and 5, respectively. By appreciatively understanding all perspectives and maintaining a positive bias towards all information, we can use the simple tools of the Apollo method to find effective solutions to everyday problems every time.

---

1   Formerly known as an Apollo cause and effect chart, but with the advent of RealityCharting® software the common term being used is Realitychart.

In Chapter 1, we talked about how most people approach problems as if they are rule-based; that is, there is one right answer. Most of our problems are really event-based—they have a number of possible answers, such as what to have for breakfast or how to prevent an accident. Our day-to-day event-based world has multiple answers to any defined problem. Based on what we learned in Chapter 2 there can be many different causal relationships behind the endless problems challenging our knowledge and skills. Event-based problems occur from an interaction of conditions and actions at a particular place and time, such as an electrical surge that led to equipment failure, the slippery sidewalk, and resulting broken leg, etc.

Another way to look at problem solving is to divide problems into those that have already happened and those that will or may happen in the future. Working on problems that have already happened is called reactive problem solving. Solving problems that may occur in the future is called proactive problem solving.

Interestingly, problems of the future are rooted in our understanding of the past. A desire to make a good living may be born out of experiencing poverty. Or, while my stomach may be full now, the goal of putting food in my mouth is born out of a past experience of hunger. When considering solutions for future problems, we first need to understand the causes of the problem. These causes are often based in experience. If we have no experience, then we must somehow attain it if we are to solve the problem. Furthering our knowledge by formal education can provide the needed experience. Using another person's knowledge will also work. RealityCharting® can help by creating a basic fault tree that identifies what could possibly cause a given event. From this analysis you can identify which causes need to be controlled to cause the desired event to happen. While this is not the normal usage for RealityCharting® it has been used this way.

Most problem solving is reactive and is the primary focus of this book, so let's look closer at event-based problems that have already occurred. In daily events a problem is generally an effect of consequence that we want to alter or prevent from recurring. To change the effect, we need a solution. In its simplest form, event-based problem solving identifies a solution that attacks the effect of consequence. Graphically, it might look like Figure 3.1.

Figure 3.1. Event-Based Problem

For the solution to prevent recurrence or accomplish our goals, we rely on our experience of the situation, intuition, guessing, and faith. Sometimes our solutions are effective and sometimes they aren't. When we fail, we shrug our shoulders and say things like "stuff happens," "it was meant to happen," or "oh well." We fail to understand that because everything has a cause and problem solving is about controlling causes, we need to identify the causal relationships between the solution and the primary effect. Therefore, an effective problem-solving process must look something like Figure 3.2.

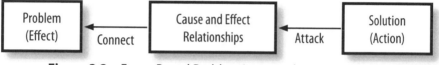

Figure 3.2. Event-Based Problem Improved

The solution attacks one or more causes in some causal set to prevent the problem from occurring. That is, solutions require an action that acts upon some cause in a chain of causes such that the effect of consequence (Problem) does what we want it to. For example, if the problem is a fire and our goal is to prevent the fire, then we can look at the causes of the fire and identify an action that will prevent it from occurring. The causal relationship for a fire would look something like Figure 3.3. If I choose a solution that acts upon all ignition sources by removing them, there will be no fire. Effective solutions, then, act upon causes such that the problem can no longer occur. Without knowing the causal relationships between the problem and the solution, we cannot expect to find effective solutions, except by intuition and chance.

Effective problem solving, then, is the process we go through every time we solve a problem of any kind; mechanic, computer engineer, doctor, lawyer, or chief executive officer all follow this principle. They differ only by the specific causes and their relationships. It is fundamental and it is based on the cause and effect principle, which we discussed in Chapter 2.

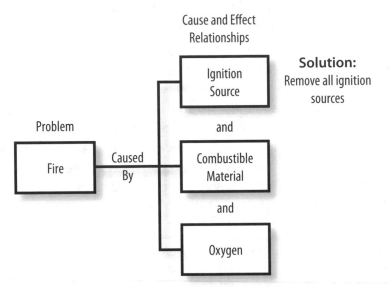

Figure 3.3. Solutions and Causal Relationships

> **Effective problem** solving is identifying causal relationships and controlling one or more of the causes to affect the problem in a way that meets our goals and objectives.

Effective Problem Solving

Effective problem solving involves two key elements: appreciative understanding and an effective tool set. These two elements must work hand in hand if we are to be effective problem solvers. Appreciative understanding is the ability to evaluate incoming information with suspended judgment until its value is determined. By using the Apollo tool set, we can establish the value of each stakeholder's perspective as it relates to the problem. This tool set also allows us to be humble in the presence of the infinite set of causes that is *Reality*.

Appreciative Understanding

"Where's my husband?" Sally asked the maitre d'. "He said he would meet me here." The maitre d', who didn't know Sally or her husband, looked puzzled, thinking, "How does this woman expect me to know where her husband is?" Indeed, why would Sally assume the maitre d' would know where her husband was? We tend to operate on the assumption that if I know something very well, then it is obvious and you know it, too. The maitre d', while wanting to be a good host, rapidly forms an opinion of Sally, and it is not as an intellectual.

This situation and others like it occur every day. When I travel, I am often party to discussions about the local football team. Others assume I know who all the players are. I don't, yet the discussion goes on. I am polite and only bring this to my host's attention if it will make a difference.

When others assume that we are on the same wavelength and we are not, we make a judgment about them. Judging what others think involves comparing paradigms. If our realities clash, the value of the other person's thoughts are given a low score and typically discarded. If the discussion is important, we often seek to convince the other person how valuable our thoughts really are.

To be effective problem solvers, we must first understand that no two people share the same reality. We must also learn that this is an advantage because it means that every person can bring a different perspective to the table. Each perspective provides pieces to the giant puzzle of life. Diversity of thought often contains the gold nuggets that make our solutions shine.

When working with other people to solve problems, we must suspend judgment on their perspectives until more is known. If we can't suspend judgment, we must think positively about what we hear. The value of everyone's ideas will be determined later by all participants in the problem-solving process. Since we seem to be designed to make judgments about others, we may need tools that will help us suspend our judgment.

Throughout the problem-solving process, we must be keenly alert to misplaced judgments. Accept all information at face value until the picture becomes clearer. The Realitychart will ultimately determine the value of all suggested causes. If a cause fits properly with other causes to form an elemental causal set, it has value. If a cause has evidence to support its existence, it has even greater value. All other suggested causes will linger and fall off the chart because they simply won't belong. This process occurs naturally and elegantly as part of the discussion, with no need to argue, make a point, or defend a position. Each cause is appreciated for the value it brings to the common reality taking shape in front of and with the participation of each person.

Each stage of the problem-solving process requires appreciative understanding. During problem definition, the accuracy of the problem statement depends on a genuine concern for appreciating everyone's perspective. Dismissing comments as having no value at this stage usually stops people from talking. When someone offers their thoughts, they believe they have value. To discard their ideas is to imply the person has no value.

Through open and honest dialog we often discover that what we thought was outside our control is actually well within our control; we just needed the courage to take action. That courage is often realized in group discussions, where we gain a better understanding and confidence from others. With appreciative understanding comes the realization that diversity of thought holds the greatest value in an organization. Until now, this has been nearly impossible because we have not had the communication tools necessary to overcome our argumentative nature. With the Apollo tool set, any perspective is allowed, but only those with obvious value will remain; and that value will be easily determined by the nature of the process.

Effective Tool Set

The Apollo tools are carefully designed to overcome each of the causes of ineffective problem solving, which were discussed in Chapter 1: incomplete problem definition, unknown causal relationships, and a focus on solutions. The Apollo method has four phases:

1. Define the problem.
2. Create a Realitychart using RealityCharting® software.

3. Identify effective solutions.
4. Implement the best solutions.

This chapter focuses on the first two phases. Phases 3 and 4 are covered in Chapters 4 and 5, respectively.

Using the Apollo problem-solving tools is quite simple and can be accomplished by almost everyone. Appreciative understanding is a little harder because it requires a change in the way we think. Fortunately, the Apollo tools help here also. By using the Apollo tools discussed in this chapter, the user begins to see how easy it is to accommodate all perspectives without the usual debate. Once learned, it becomes a habit for life, to be used in all communications.

Phase 1: Define the Problem

The first step in the tool set is defining the problem. Let's look more closely at a typical business meeting where we have to decide how to resolve a pressing issue. The scenario goes something like this.

Boss: "OK, I'm glad everyone could be here. As you know, we have experienced another [insert your problem here]. This is the third time this has happened in as many weeks. Now, we all know what's going on, so what I want to do here today is fix this thing once and for all. Does anyone have any ideas on how we can prevent this from happening again?"

Wendy C. Mitchell

As the boss makes the statement about everyone knowing what the problem is, many people are nodding their heads in agreement, giving him the positive feedback that they are in tune with his thinking. They may even smile at one another in agreement. The body language is ripe with consensus and the air is full of confidence, but problem definition is not discussed. The stated goal is preventing recurrence and solutions are solicited. Discussion ensues with expressed opinions and everyone carefully listening to the boss to know which way to lean. Yet everyone sees and expresses the problem differently. Few, if any, of these team members will see the problem being within their domain. They will see the problem from their perspective, but their typical solution will be to have someone else change.

The dynamics of these decision-making meetings are predictable, and the path is the same almost every time. When we look more closely, this is what happens:

First, the problem is not defined. It is assumed that everyone knows the problem. After all, it has happened at least three times, so everyone surely knows what it is by now. The assumption is that it is so obvious we need not waste time discussing it. This thinking is based on the illusion of common sense. In the meantime, every person in the meeting is thinking to themselves, "Yes, I know what the problem is — if the other manager would only get off the dime and do what I told him to do, this wouldn't be happening." Each person sees a piece of the puzzle based on their perspective, but none see the whole picture. Eventually, communication stops, the boss is left to make a decision, and no one is happy.

Second, there is no discussion of causes; or if there is, a few causes are expressed using storytelling. Any discussion about causes usually deteriorates into a debate won by the person who can tell the best story, usually the boss.

Third, the discussion is centered around possible solutions. We are so solution oriented, we ignore the causes and debate the solutions. The analysis phase of problem solving is essentially ignored because we mistake decision making for analysis. Managers especially slip into this trap. They listen to various stories and see their role as the decision maker, not the analyst. They use their "experience" as the basis for the decision when they should be using the known causal relationships. My experience has shown that once managers see the Apollo problem-solving process and how well it presents the problem, they are not satisfied with anything less.

Defining the problem should seem obvious to us, yet we fail to do this adequately about 95% of the time. Sure, we state something as being wrong or bad or unacceptable, but we don't stop and write it down or take the time to fully understand the significance of the problem. The act of writing down the what, when, where, and significance of the problem provides focus. We fail to do this because the problem seems so obvious: the plant shutdown, or the lost-time injury, or the poor quality of service. These things have happened before so we don't have to define them, or so the misguided thinking goes.

Defining the problem is the first step in the problem-solving process, so take the time to do it adequately. The following discussion provides detailed guidance to help you be successful. In the Apollo process, getting the "right" problem definition the first time is not as critical as other methods because as you identify more causes, you may find that you started in the middle and need to redefine the problem. Quite often the problem is something bigger than you originally thought and you need to redefine it. This is very easy to do with the Apollo process, so don't waste a lot of time trying to make sure you have a perfect problem definition the first time.

Complete Problem Definition

A complete problem definition should include four elements:

- *What is the problem?*
- *When did it happen?*
- *Where did it happen?*
- *What is the significance of the problem?*

The what of any problem is the effect of consequence. This is the effect we do not want to recur, and hence we are calling it the primary effect (as discussed in Chapter 2). The primary effect is the beginning of asking "why." It is a noun-verb statement such as "clock stopped," "arm broke," "system failed," etc.

The when of any problem is the relative time of the primary effect. It may be the time of day or the point in a sequence of causes, such as "after the clock fell."

The where of any problem is the relative location of the primary effect. It may be the physical coordinates on a map or the position relative to something else, such as "the swimming pool next to the tennis courts."

The significance of any problem is the relative value the primary effect has on you or your organization. It is the answer to the question, "Why am I (we) working on this problem?"

When we put all these together, we have a clear understanding of the problem. Continuing with our clock example, we could have the following problem definition:

> **What:** Clock stopped
> **When:** 10:32, after it fell
> **Where:** The swimming pool by the tennis courts
> **Significance:** The swimmers cannot train effectively.

Knowing the significance upfront not only helps us prioritize the need to work on the problem, it also helps us determine which causes to pursue and which solutions are within our control. Significance can involve many factors, but the most common ones are cost, safety, and frequency.

Properly assessing the significance of the problem is perhaps the most important element of defining the problem. By knowing the significance in the initial problem definition phase, we determine the required effort and priority of the problem before proceeding; we may even determine not to pursue the problem at all.

If a problem resulted in the loss of business or a severe injury, the problem is significant and deserves attention. In personnel performance issues, frequency may determine our corrective action. If an employee who frequently makes mistakes caused a problem and our causes lead us to find he/she chooses not to learn, our corrective action legitimately could be to terminate employment. But, if the problem was caused by the lapse in concentration of a valued employee after 20 years of error-free work, the significance is radically different; and we may choose to do nothing for the exact same set of causes.

As another example, if a fall caused a broken wrist and the wrist is mine, the significance might be minimal if it happened once in 30 years of snow skiing because the frequency is acceptable to me. With this minimal significance, I may only need to understand a few causes to help me avoid the conditions that set me up to fail. If, on the other hand, the broken wrist was that of a professional football player, this was the third time it has happened, and it kept three players off the field for six weeks each, then the significance is much greater than my skiing injury. In this situation we may need to understand a hundred causes to find an effective solution. Each situation is different, and you will learn from experience how far to go to find the best solutions.

The greater the significance, the more important it may be to know the causes between the causes because each new cause adds more opportunities for an effective solution. Including the significance in the problem definition is essential to effective problem solving.

When stating the significance, be specific and try to avoid categorical statements. Instead of injury, the significance may be stated as "lost use of hand"; or, instead of "plant shutdown," we should state "lost $50,000 in production costs."

Significance is relative to our goals and objectives. If our organization has a goal to produce something safely and economically or to provide the best service possible, the significance of the problem should center around these goals. Oftentimes organizations fail to communicate their goals and objectives to each employee. When employees do not know what their specific goals are, they find it difficult to identify the significance of a particular problem. The result is an incomplete problem definition and difficulty in determining an effective solution. The employees are often left trying to guess what the boss wants, rather than thinking for themselves to accomplish an important objective.

Knowing the significance also helps us know which questions to ask during the analysis phase. As we go down the various cause paths, it helps us decide which paths may provide the best solutions. For example, if my problem is a waste spill, then knowing what kind of waste it is not only tells me what the significance is, it also tells me which "why" questions to ask later. If it is just dirty water, then I may only need to find out why it got out of its container. If it is toxic waste, then I may need to ask more "why" questions to find out why it was not contained by a secondary containment system.

Significance is completely relative and unique to each problem, but if we don't define it upfront, it has a tremendous negative impact on our ability to effectively and efficiently solve our problems.

What the Problem Definition Is Not

You will notice that the problem definition does not ask "who" or "why." "Who" questions lead to placing blame and are generally a waste of time. "Why" questions are reserved for the analysis phase of problem solving.

The only acceptable "who" question is, "Who knows the answer to my other questions?" Unfortunately, "who" is one of our most often asked questions, and we need to stop doing it. Asking "who" is understandable, not only because we seek to know who caused the action but also because of a very personal human condition. Consider this: When we are presented with a failure of consequence that we feel personally responsible for, the very first question we ask ourselves is, "Did I screw up?" If the answer to this question is "no," we immediately and universally ask the next question: "Who did?"

If the answer is "yes, I screwed up," we either accept responsibility and try to learn from our mistake or more often seek to find a way to implicate other people or other things. If the answer is "maybe I screwed up," we seek to divert negative consequences by developing a rationale that points elsewhere: "The devil made me do it."

The point is, we need to understand that this need to find a "who" is ingrained in our being and is therefore very difficult to stop. It must be a conscious effort. Asking someone to remind us of this tendency often works well because it keeps both people more conscious of the need. In time it becomes a habit and we stop asking "who"; and when we catch others doing it, we remind them not to go there.

The "why" questions may be asked early in the problem-solving process. While this is not wrong, it can be inefficient. It should be held back until the problem is defined and written down.

Because causes and effects are the same thing and we are seeking the primary effect in the what of the problem definition, we will inevitably get into causes during the problem definition phase. However, asking "why" is the essence of the analysis phase and is a separate step unto itself. It is not that we have to define the problem and leave it never to come back, but we need a place to start. By defining the problem, we define the starting point and can get on with asking "why."

It is common to start with one primary effect only to realize that it is a symptom of a more significant problem and that we need to redefine the problem. As we discover the different perspectives of each contributing person (stakeholder), several elemental causal sets will be generated. They should be noted, but the focus should remain on finding the one effect that seems to stand out as the most significant one. The causal sets or individual causes should be documented for later discussion.

Conflicting Goals

Defining the problem is usually very easy, but when many stakeholders are involved, such as public projects, a new dynamic presents itself and we need to be aware of it. The purpose of event-type problem solving is to learn from history and prevent the problem from ever happening again. Unfortunately, this is not always the case when politics are involved. One stakeholder's goals may be different than another or from those of an organization. Because goals define the purpose and purpose defines the significance and significance helps define the problem, conflicting goals will result in an impossible problem-solving environment.

For example, in the Northwestern United States there is one political goal of saving the salmon in the rivers and the proposed solution is to remove all the hydroelectric dams on the Columbia and Snake Rivers. Another political goal is to provide a viable economy by providing carbon-free electricity for the region and irrigation water for farms to grow food. Unfortunately, the proponents of dam removal have supplanted their goal of saving the fish with a solution to remove the dams. Without any comprehensive causal analysis by anyone they have determined that dam removal is the only solution to meet their goal. With this conclusion, they have changed their goal from "Restore salmon runs" to "Remove dams." They have moved from focusing on the undesired effect of decreased salmon to a solution mindset that no longer incorporates causal relationships. Since this is in direct conflict with the goal of providing a viable economy and food to eat, the two entities have continued this loosing battle for over 20 years and until such time as politicians realign their goals and do a comprehensive causal analysis, the folly will continue.

Phase 2: Create a Realitychart

Based on the cause and effect principle of Chapter 2 we are able to create a simple set of tools that will allow us to establish a common reality from the diversity of thought in any organization. I've broken the Phase 2 discussion into two parts: the Principles and the Mechanics of creating a Realitychart.

Principles of Creating a Realitychart

Creating a Realitychart has five elements or steps:

1. For each primary effect, ask "why."
2. Look for causes in actions and conditions.
3. Connect all causes with "Caused By."
4. Support all causes with evidence.
5. End each cause path with a "?" or a reason for stopping.

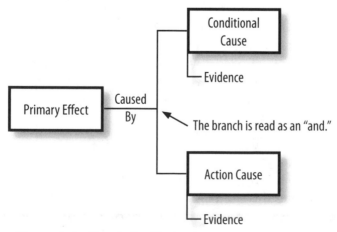

Figure 3.4. Simple Realitychart

We start with the "What" we identified in the initial problem definition. This is called the "Primary Effect" and is the point at which we begin asking "why." We then identify the conditional and action causes of the primary effect. The third step mitigates the detrimental effects of storytelling by connecting all causes with the words "Caused By." The fourth step adds evidence to support the existence of stated causes. Adding evidence makes groovenation and denial difficult to maintain. The fifth element prevents us from stopping too soon by forcing us to address why we stopped. These fundamental elements of the Realitychart are found in Figure 3.4.

Step 1: For each primary effect ask "why." When asking "why," look for causes in the form of actions and conditions. When starting the problem-solving process, there is usually not enough information to feel comfortable about the primary effect, so you may have several, in which case you'll need several charts. Instead of arguing or wasting time trying to decide on one primary effect or starting point, list the primary effects you have and attack each one—one at a time—by looking for the connecting causes. What is happening during this stage of the process is that each team member is expressing their own reality about what happened, and they often differ. No one is right or wrong; these are just different perspectives being presented. The purpose of this process is to capture all causes and their relationships.

As previously stated, the process is much like putting together a jigsaw puzzle—we start with one piece and try to find a match. We put together a few pieces at a time and then run out of connections, so we start with another piece and repeat the process. With each success we create an elemental causal set (as defined in Chapter 2, a single causal relationship made up of at least one conditional cause and one action cause, to create an effect). Eventually these elemental causal sets will combine to form a picture or common reality just like a jigsaw puzzle. If the pieces don't fit, they are probably part of another problem or are inconsequential data. The only difference between creating a jigsaw puzzle and the common reality of the Realitychart is the chart has no boundaries.

This concept of no boundaries is difficult to get used to because there is a strong sense that you have lost control. The fact is, you never were in control so don't let these feelings get to you!

Nothing seems to fit together, so keep reminding yourself of the jigsaw puzzle; it will always fit together if you have the perseverance to follow these strategies. Have faith in the process; and with a little patience and tenacity, the picture will become clearer each time you repeat the steps.

When the Realitychart is finished, there is usually only one primary effect left. If you have some causes left over, chances are they are part of another problem.

The more complicated the event, the greater the chance of multiple primary effects. Remember, because of the cause and effect principle, all causes are connected with all other causes in some way. Trying to show these connections may make your understanding of the problem too complicated. If the problem is as big as an elephant, how should we eat the elephant? One bite at a time, of course. Eating it whole would be impossible, so we need to divide the problem into smaller parts by focusing on separate primary effects. We call this chunking and it allows each part of a problem to be assigned to different teams.

Step 2: Look for causes in actions and conditions. As the causes present themselves, do not be concerned with whether they are actions or conditions. Rather, just concentrate on identifying the causes. After you've listed the known causes, go back through the cause and effect chart and look for branches. If you have written

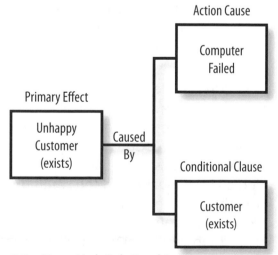

Figure 3.5. Noun-Verb Relationships

down an action-type cause, ask yourself what conditions had to be in place for the action to cause the effect. For each conditional cause, make sure you have a corresponding action cause. Remember, we generally find several conditional causes and one action cause in each elemental causal set.

The only value of knowing if a cause is an action or a condition is that it tells us which one is missing and hence which one we need to look for. In a completed cause and effect chart, there is no value in knowing the difference between an action cause and a conditional cause.

When we express a cause or an effect, we see that it has a name and an action. The "what" of each cause/effect is stated in a noun-verb or verb-noun expression. In the case of a conditional cause, the verb is often understood as "exists" or "is." For an action cause, the verb is the action and the noun is the thing that is acting or being acted upon.

In Figure 3.5, the primary effect is expressed as a modified noun (Unhappy Customer). The verb is understood as "exists." The action cause is expressed as a noun (Computer) and verb (Failed). The conditional cause is expressed as a noun (Customer). Again, the associated verb is understood to be "exists." When expressing cause and effect relationships, we should always attempt to follow these conventions. While we should not waste time checking every cause to ensure compliance, following this convention will improve clarity.

Action causes are more readily identified than conditional causes. As an example, consider the sequence of events below:

Time	Timeline Entries
0935	High-temperature alarm sounded on the outboard bearing of pump 102.
0945	Fire alarm sounded in Building 603.
0946	Operator reported fire; fire brigade was activated.
0947	Plant manager was phoned at home.
0949	Fire suppression system was activated in Building 603.
0950	Catalyst tank exploded.
1002	Fire alarm sounded in Building 604.
1010	Building 604 exploded.
1011	Fire suppression system was activated in Building 604.

Every one of these timeline entries is an action cause. By focusing on actions and not on the associated conditional causes, we leave out important causes that might be acted on to provide an effective solution. If we add some effects and conditional causes to this picture, the elemental causal sets begin to emerge. In Figure 3.6, we can start with the action cause of high temperature provided by the timeline and build on it by adding the effect and a conditional cause.

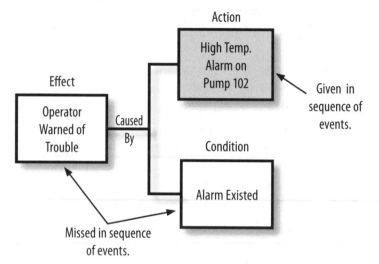

Figure 3.6. Effect of Adding Conditional Causes

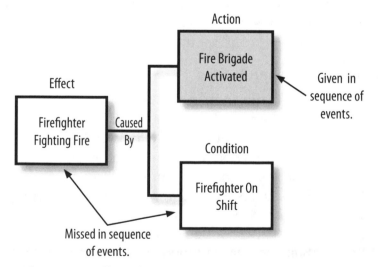

Figure 3.7. Effect of Adding Conditional Causes

In Figure 3.7 another elemental causal set is identified. By evaluating each action cause in the timeline, we are able to create some elemental causal sets that will eventually come together to form a complete cause and effect chart.

In another elemental causal set (Figure 3.8), we find a conditional cause of "building existed," which seems fairly innocent at first glance, but moving on down the cause path we find that the building should have been torn down. If this had happened, it would not have exploded, so we may have discovered a programmatic cause regarding planning and scheduling problems. Each time we develop an elemental causal set, we enrich our understanding of the problem and increase the number of causes, each of which may provide an opportunity for prevention.

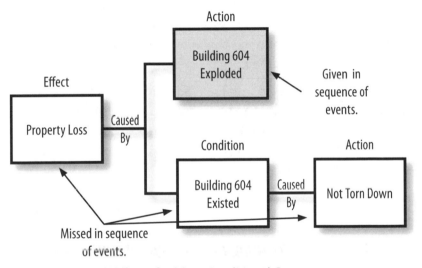

Figure 3.8. Effect of Adding Conditional Causes

The best solutions are usually associated with conditional causes, partly because of our greater ability to control conditional causes, whereas people or action causes are less predictable.

It is often advantageous to start with a sequence of events and use it to build your causal set. It provides a good skeleton by identifying the action causes that seem important to the event. Be cautioned, however, that some event entries from a sequence of events can be meaningless when put into a cause and effect chart because they may be story elements, not causes. In the previous example, the event entry, "0947, Plant manager was phoned at home," is of little value as a cause because it is a story element, not a cause. Story elements are important to a story because they have implied meanings, but they do not provide causal information.

Sometimes causes are non-causes. That is, they are non-actions or non-conditions. For example, the last action in Figure 3.8 states that the building was not torn down. While a negative statement, it is treated as an action cause. The same can occur with conditions. We can have "no firefighters on duty" as a condition.

Step 3: Connect all causes with the words "Caused By." Using "Caused By" helps the mind to align causes from the present to the past. Because we are starting with an effect we do not want to recur, we must understand its causes that lie in the past. This tool ensures we follow the fourth element of the cause and effect principle: an effect exists only if its causes exist at the same point in time and space. An additional benefit is that it avoids storytelling if we follow what I call the Square One Loop.

> The Square One Loop involves following each cause path in a Realitychart until the collective point of ignorance is reached, and then starting over again with the primary effect (square one) and repeating the process.

The key to efficiency in the Apollo method is the Square One Loop. It works like this: As you ask "why," immediately write down the answer, and ask "why" again. If you are in a group or team meeting, minimize discussion by writing down the first cause you hear and immediately asking "why," thus cutting off further discussion (storytelling) and over-analysis. Continue to ask "why" until the answers stop coming, called the point of ignorance, or things get fuzzy, called the fuzzy zone. This is where you can honestly say, "I don't know and neither does anyone else on the team."

Follow each cause path to the point of ignorance; and when you reach your collective point of ignorance, go back to square one (the primary effect) and start asking "why" again (this represents one loop in the Square One Loop). Try to identify causes between the causes and look for branches (actions and conditions) each time you go through the loop. Repeat the Square One Loop several times or until the ends of every cause path become fuzzy—your point of ignorance. If you want to make assumptions or express opinions, do so within reason, but use a "?" to reflect your doubts.

Focusing on "caused by" as the connecting phrase will minimize unnecessary analysis and storytelling. As you go through the Square One Loop, say out loud, "The (effect) was caused by" with no inflection. Saying "caused by" with a question in your voice elicits a narrow response because it implies one answer. If you are not getting discussion, ask "why is this cause here," or state the effect and simply ask "why." Saying "why" with no inflection also elicits a broader response because it implies no boundaries.

Step 4: Support all causes with evidence or use a question mark. A unique and critical element of the Apollo problem-solving method is incorporating evidence. Each cause must be known, and the evidence of its existence must be documented on the Realitychart. At some point, as you go through the Square One Loop, usually the third or fourth time, begin to add evidence to each cause. In the Apollo method, evidence is written below each cause box.

For each unanswered "why" question and each cause that does not have supporting evidence (represented with a question mark), assign someone to find more information. RealityCharting® software automatically creates an Action Items List for every question mark on the chart. Set a time to reconvene the meeting or finish the Realitychart, and input the results of your findings. Repeat this process until you are satisfied with the results of the Realitychart. This is not an invitation to over-analyze or study the problem to death. Rather, it is a desire to incorporate as much knowledge as reasonably possible given your time constraints. It usually takes only one iteration of going outside the group to find additional evidence and unknown causes.

The question often arises, "How do we know we have good evidence?" We seem to have an innate understanding of what evidence is. I find very few people have difficulty establishing causal evidence where it is available. However, when asked to define evidence or explain what makes good evidence, most people can't do it.

The dictionary defines evidence as data that supports a conclusion. We conclude something exists either by directly sensing it with one of our five senses or by inference through causal relationships. We also use intuition and feeling as the basis of conclusions, which are more subtle forms of inference.

Sensed evidence is the highest quality of evidence and consists of knowing by way of sight, sound, smell, touch and taste. In the example in Figure 3.9, all evidence is sensory: it was seen or heard. Evidence is best stated by telling which sense was used. If something was observed, we know through our eyes; if we smelled smoke, then we know through our nose.

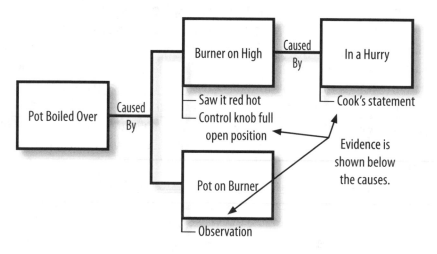

Figure 3.9. Sensed Evidence

Inferred evidence is known by repeatable causal relationships. Evidence that someone is happy can be known by a smile on their face. That is, we infer happiness by knowing the following repeatable causal relationship: Smile caused by Facial Muscles Moving caused by Happiness.

The best evidence is direct observation using one or more of our senses. For example, I know fire because I see flames, feel heat, and hear crackling. Inferred evidence is less desirable but may be all we have. Using this same example, we may know fire by seeing smoke, smelling smoke, and tasting smoke, but this is an inferred causal relationship that assumes fire causes smoke.

Causes and evidence are often interchangeable because of the way we use inferred evidence. I may legitimately state that smoke is evidence of fire. But it is also correct to say smoke is caused by fire, and the evidence of the fire is my observation of smoke. If I can't see the flames of the fire, using smoke as inferred evidence may be acceptable, but it is a lower quality of evidence. The perceived smoke may actually be mist or fog and there is no fire.

Because inferred causal relationships are not always well understood, they are not necessarily as reliable as sensed evidence. For example, if I believe that wind is caused by clouds, as I did as a child, then it is logically inferred that big clouds are evidence of high winds. If this causal relationship helps me understand my world and is repeatable within that world, then I will continue to infer that big clouds are evidence of high winds, even though it is scientifically false. We can only know what we know.

If inferred evidence is all we have, we should use it. Sometimes the only way to know something is by inference. For example, we cannot directly sense the pressure in a tank because we don't have built-in pressure sensors. We may sense pressure if we let it out and it impinges on our skin; but while it is in the tank, we can only know the pressure indirectly through inferred causal relationships. Reading a pressure gauge may be evidence of high pressure. We know the mechanics and the physics that cause the pressure gauge to work, so the pressure is known through a repeatable causal relationship of the instrument. This is shown in Figure 3.10. The pressure indicator reading is caused by the indicator needle moving, which is caused by the bellows expanding, which is caused by high pressure. Because this causal path is known and repeatable, we do not need to write it out. "Pressure Indicator Reading" is adequate evidence of high pressure if everyone who reads the Realitychart knows this relationship is valid and repeatable. If it is not understood by everyone or is not repeatable, the cause path should be explored to verify that the causes exist with sensed evidence.

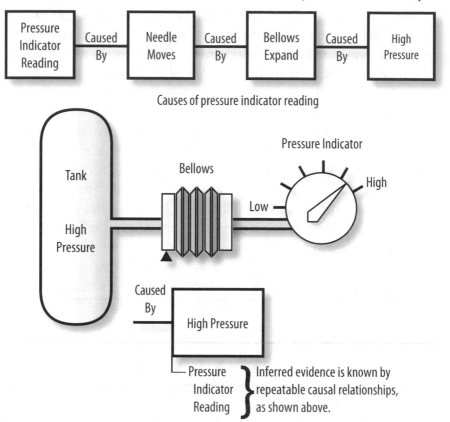

Figure 3.10. Inferred Evidence

Because inferred evidence relies on the assumption that the reader knows the causal relationship, it should be readily verifiable. If it is not, the causal relationship should be included in the Realitychart. For example, consider the inferred evidence in Figure 3.11.

"Fuel Vaporized" is evidenced by the temperature of 285°F, and this can be readily verified by looking at a fuel vaporization chart. "High Tank Temperature" is evidenced by reading the temperature recorder. The reading of the recorder is sensory evidence of inferred evidence. The instrumentation that causes the recorder to work is known by repeatable causal relationships. But what about "Spark" being evidenced by "Electric Motor?" The inference is that because electric motors can make sparks, the electric motor is evidence of a spark. But this is not necessarily so because not all motors make

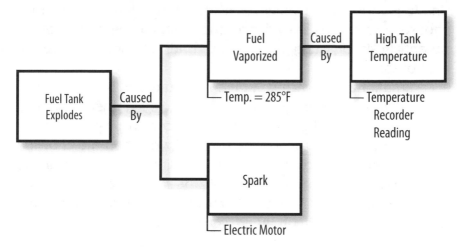

Figure 3.11. Inferred Evidence

sparks. If we knew it, it would be better to indicate evidence of a spark as "burn mark" or "saw electric arc" and show the spark as caused by "Electric Motor." The point is that when using inferred evidence, be very careful that you know the causal relationships being expressed with the inference.

Intuition is inferred evidence based on both reason and emotions; but because it occurs at a subconscious level, we are not capable of explaining where it comes from. Consequently, using intuition as evidence presents a risk. An example of intuited evidence is shown in Figure 3.12. It is understandable that

Figure 3.12. Intuited Evidence

"Inadequate Marketing" could be evidenced by "Intuition." We may choose to accept this as evidence, but we should be suspect of the potential risks of acting on this cause until we know more causal relationships.

Emotional evidence is shown in Figure 3.13 where "Sensed Danger" is evidenced by feeling scared. We see that emotional evidence is inferred evidence from a known repeatable causal relationship, but the five senses are not involved in the knowing process. Emotions and feelings exist in the limbic system within the old reptilian portion of the brain, while the senses are located in the cortex along with reasoning. As such, emotions and reasoning are not well connected. Emotions are very real and they should not be ignored as evidence of a cause, but they should be held suspect because they are not always reliable.

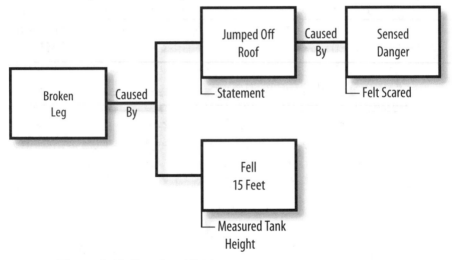

Figure 3.13. Emotional Evidence

Once we leave sensed and inferred evidence, we enter the realm of risk very quickly. Providing evidence is very important to the process, but what I have concluded after many years of evaluating Realitycharts is that the quality of evidence does not necessarily cause poor solutions. It can prevent you from finding the best solutions, but it won't prevent you from finding effective solutions. Real-world Realitycharts with little or no evidence often lead to solutions that are far better than the ones found through storytelling and categorization schemes. Cause and effect thinking is incredibly powerful. Even if the chart has not been constructed properly, the thinking behind it allows the user to be more effective than ever before because our other strategies are so ineffective.

While most people seem to intuitively understand what constitutes evidence and what does not, it is not universally known. Sometimes certain words get in our way. The most misused word I've encountered is "fact." For most people a fact is something that is absolutely known to exist in their world and yours. The problem with this notion is that it ignores perception, as discussed in Chapter 1. What may be a "fact" to you may inspire a great debate in your neighbor. To avoid this issue, I suggest you never use this word or redefine it to include evidence.

> Fact: A cause supported by evidence.

Facts have no value unless used in the realm of causal relationships. It is a fact that the sign is red because we can see it, but this statement has absolutely no value whatsoever until such

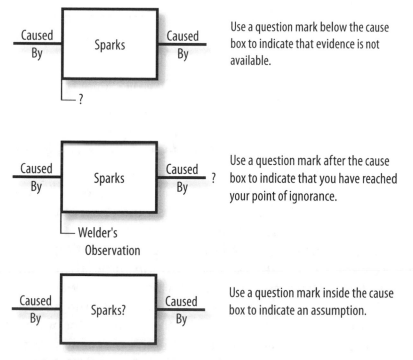

Figure 3.14. Use of Question Marks

time as it is placed into a causal relationship. The red sign caused me to stop or "Stopped Moving caused by Red Sign." Therefore, meaningful facts are always evidence-based causes.

Sometimes we find contradictory causes and evidence. Contradictory causes should be documented on the Realitychart and evidence sought to explain the contradiction or establish one cause as more likely than the other by virtue of preponderance (such as weight, quantity, or importance). The Realitychart works well to accommodate contradictions because it lays out all perspectives. Every stakeholder can see the relationships between other causes and supporting evidence. If one cause has a preponderance of evidence and a competing cause has poor evidence, the chart allows both to be represented with little or no debate. If a solution is attached to a cause chain with poorly evidenced causes, then it clearly shows the risk being taken. If the solution is attached to a cause chain with well-evidenced causes, the effectiveness will be assured.

Often, the problem with evidence is not being able to find it. If you cannot find evidence or cannot find a cause, use a question mark in the ways shown in Figure 3.14. These question marks denote that an unknown exists and hence represents a risk if a solution is attached to this cause or the cause chain it is in.

Step 5: End each cause path with a "?" or a reason for stopping. Our purpose in asking why is to find our point of ignorance, not show how smart we are. We need to embrace our ignorance, which should be quite easy now that we understand the notion of the infinite set of causes that is *Reality*. When you get to your point of ignorance, put a question mark to denote your lack of knowledge and the need to get more information. RealityCharting® allows you to insert a "?" with the click of the mouse and it will automatically put the associated cause into an Action Items List.

Only one in twenty people are capable of admitting they don't know when dealing with a serious question within their expertise. About 95% of the time, when we can't find an answer, we make one up and then spend enormous amounts of time justifying it with various rationalizations. Because this is a fundamental human condition that is very detrimental to effective problem-solving, the

Apollo method requires it be addressed and provides an easy way to document what you don't know in RealityCharting®.

Sometimes a cause path will take us not to our point of ignorance, but to a valid reason for stopping. There are really only four reasons for stopping and RealityCharting® again makes it easy to insert these with the click of a mouse:

Desired Condition – This is the most common reason for stopping and reflects the fact that the event was caused by the pursuit of one or more goals. If you reached your goal or a desired condition, there is no need to continue asking why. If your goal is faulty, that is another matter. Examples of this include "met productions goals," "procedure followed," or "service level met."

Don't Have Control – This can be an easy excuse for stopping, so make sure the lack of control is outside you or your organizations control before using this reason. Examples are "laws of physics" or "legal requirement."

New Primary Effect – This occurs when you get to a point in the cause path that you need to do a separate analysis. The reasons for a separate analysis can be many, such as outside your control, but within another organizations control, or it is within your control, but you want to separate it for resource or presentation purposes. This reason is often used as an interim stopping point because you have a separate team working on the details of this cause. When they are done, you can easily import their Realitychart into the final analysis.

Other Cause Paths More Productive – Sometimes you may have solutions that will prevent recurrence and there is no need to go down other cause paths because they are simply not productive or cost too much to pursue. It doesn't make any sense to continue down these cause paths and spend more money when you already have effective solutions. This reason for stopping is usually not identified until late in the analysis. This may also be caused by obviously frivolous causes, like sky is blue.

Feedback Loops

Sometimes the cause path does not end, but loops back to previously identified causes. In the world of systems this is called a *Feedback Loop* and adds another dimension to the notion of an

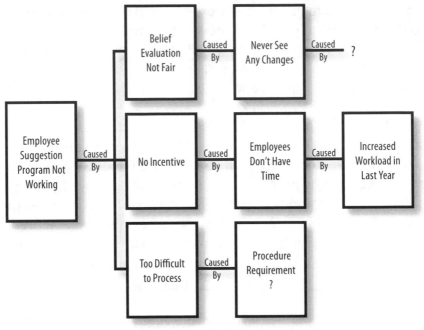

Figure 3.15. Apollo Cause and Effect Chart: First Pass
through Square One Loop

infinite set. There are two basic types of feedback loops, positive
and negative.

Positive feedback in a system is where the increase in a given
variable or cause produces a further increase in that variable or
cause. The growth in human population is a good example of a
positive feedback loop. The more people who exist the more there
are to reproduce and the numbers increase exponentially.

Negative feedback in a system is where the increase in a given
variable or cause produces a decrease in another and vise versa.
Negative feedback loops often produce stabilizing effects such as
war, pestilence, and famine, which often cause a decrease in the
human population.

Feedback loops are common in all natural and human
systems and add another dimension to the reality of the infinite
set. With each feedback loop we add a complexity to reality that
makes it difficult to comprehend and even more difficult to express
when using storytelling and categorization as our only means of
communicating. RealityCharting® provides an easy click of the

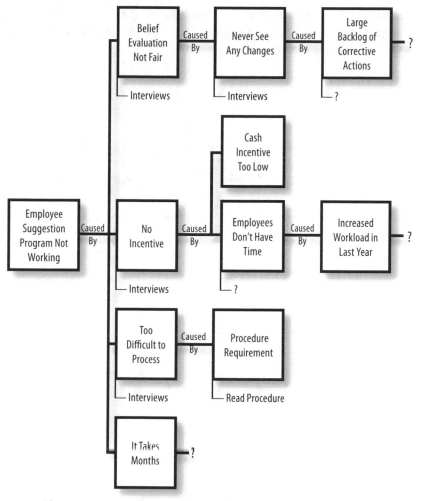

Figure 3.16. Apollo Cause and Effect Chart: Second Pass through Square One Loop

computer mouse to identify, document and share feedback loops. Something not possible in any other communication format.

Mechanics of Creating a Realitychart

While the Realitychart can be created by one person, its greatest power is realized in a team environment. When several people are sharing their realities to create a common reality, synergy is created like never before. To establish favorable conditions for the Realitychart in a team environment, assemble four or five people and a facilitator. The facilitator should first gather information from all team members (discussed in more detail in Chapter 6), define the problem, and write it down for all to see. The facilitator should then proceed to create the Realitychart.

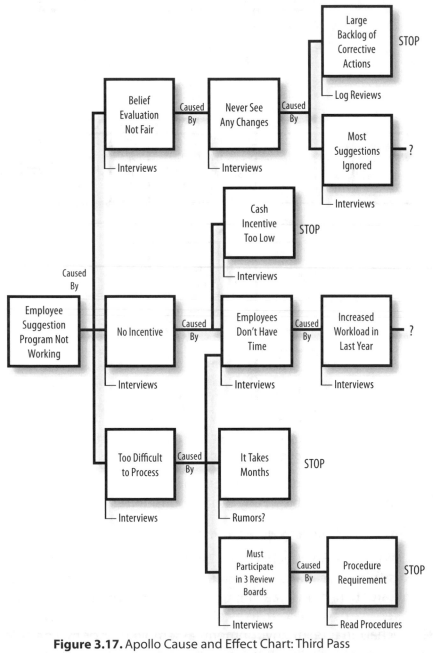

Figure 3.17. Apollo Cause and Effect Chart: Third Pass through Square One Loop

As we just discussed, creating the Realitychart is an iterative process that begins with a recognized primary effect based on the problem that was defined. As we ask "why" and look for causes, some elemental causal sets are created and then expanded into cause chains. Figure 3.15 shows the results of one trip through the Square One Loop. Notice, there is no evidence yet, some cause paths end in question marks, and there is only one three-pronged branch so far.

After going through the Square One Loop a second time, illustrated in Figure 3.16, more causes are discovered, another branch is found, and some evidence has been added. There are no rules for when to add evidence, but we have started to add it in this example. I recommend adding evidence on the third or fourth time through the Square One Loop.

As we go through the Square One Loop one more time, shown in Figure 3.17, we discover more causes, more branches, and complete the evidence. We can choose to stop now or continue. If we stop, we begin the solutions phase of the process (discussed in Chapter 4).

The following are some hints for effective charting.

Mechanics of Charting – Creating a chart requires making a draft and formalizing it for further discussion and finalization. The draft may be created manually or by using RealityCharting® software. Both methods are equally effective if the key stakeholders are involved.

Manually Created Draft: Using large self-adhesive notes, such as 3 X 5 Post-Its™, write the causes on the notes and place them on a large white board or vertical flat surface. As you go through the Square One Loop move the notes and evidence accordingly. Don't try to create the "right" chart as no such thing exists – just keep asking why and arranging the causes until you reach your point of ignorance.

Computer Software: RealityCharting® was created in July, 2002 to help facilitate the development of the Apollo cause and effect chart. Because the Apollo problem solving process is a new way of thinking it often requires great courage to use the first time and we found that many people would not use it because of their fear of embarrassment. RealityCharting® removes this barrier by providing an easy-to-use, intuitive software application that guides the user through the entire process.

Computer Software Draft: RealityCharting® provides a special feature called *Synergize* to gather initial information and helps create a draft analysis that can be shared electronically or in a printed version.

Chart Finalization: Regardless of which method was used to create the initial problem analysis, RealityCharting® should be used to document your findings because it will ensure adherence to the Apollo process described in this book and make it easier to communicate with others as you go back and forth to create the final chart. Once a draft is created, you should e-mail it to all stakeholders or team members for review and comment. As you get answers to your questions revise the chart and iterate again until done.

RealityCharting® provides a Wizard that guides the new user through each step of the Apollo problem solving process. The "What" in the Problem Definition is automatically placed on the chart as the Primary Effect. From here, the chart is easily created by the use of drop down menus. Causes are easily moved, added, or deleted by the click of a mouse. They are automatically aligned by the software and the words "Caused By" placed between causes.

A rules-check feature ensures that the causes are noun-verb statements, that each cause has evidence, that each effect has an action and conditional cause, and that every cause chain ends with a question mark or a deliberate reason for stopping. As discussed in the next Chapter, possible solutions are easily added for each cause and then checked against the solution criteria. A report is automatically generated and before being finalized is checked to make sure it meets all the requirements of the Apollo problem solving process.

Charts can be shared with anyone through the use of free RealityCharting® Reader software. Unique page layout features allow customized viewing and printing of a large chart. There is no need to cut and tape pages together because connecting pages are automatically cross-referenced on each page. RealityCharting® also allows the chart and all major report text to be exported to Microsoft® software applications by use of an *Export* feature.

Philosophies – Charting is not an exact science, so there is no single correct cause and effect chart for a given event. You are not trying to find the truth, you are trying to identify effective solutions. The cause and effect chart is simply a representation of the common reality of those who create it. If you find yourself disagreeing or arguing over the right causes, you have missed this important fact. The chart will provide a starting point from which to find effective solutions.

We are often hindered by our willingness to surrender our individuality in a group or team setting. Go outside your group to find different perspectives and thus additional causes.

Parochial thinking often causes a narrow point of view and we stop too soon in our quest for causes. Try to find other knowledgeable people to review your chart before you finish it. Listen to them with an open mind; and if you can't explain the chart, you probably don't understand the causes.

Learn to be humble and help others do the same. Be authentic and honest with others.

Effective Strategies – Look for actions and conditions, but if you don't see them, do not get bogged down. It is more important to keep moving through the Square One Loop than anything else. If you stop asking "why" or get bogged down, people lose interest and the process comes to a halt. Follow each cause path to your point of ignorance or make a conscious decision to stop asking "why." Do not discuss solutions until you are finished with the Realitychart.

When team members cannot agree or to get the team started, the facilitator can prepare a strawman Realitychart. It is always easier to criticize than to create. (A strawman is an argument set up for discussion so as to be easily refuted, like a trial balloon.)

If you ask "why" and no answers come, look for answers in people, procedures, and hardware. It also helps to look for differences; and when you find them, start asking "why." With this strategy, you will find yourself back in the cause and effect mode. During one investigation, I found myself at the end of a cause chain because when I asked the welder if he had done anything differently during the incident that led to the problem, he said he could not think of anything, yet something had changed to cause the problem of the weld not being correct. As I continued to ask questions, I found the preheating procedure was not performed properly. Once I found this difference, I could begin to ask "why" again. When something has been working for a period of time and a problem occurs, it is best to start by looking for differences or changes in the procedures.

Often, it is beneficial to start the chart by listing all possible causes. Because the causes may not have evidence at this time, the connecting logic may be "or" rather than "and." For example, the primary effect is caused by this cause, "or" this cause, "or" this cause, and so on. As you establish evidence for two or more parallel causes, the "or" logic disappears and the final logic becomes "and" (that is, the primary effect is caused by this cause, "and" this cause, "and" this cause.) If you choose to show possibilities on the final chart, they must be labeled accordingly, because all parallel cause boxes are understood to be related by "and" logic. "Or" logic means you don't know. It is caused by this "or" that. By adding evidence that the causes exist, the "or" must be changed to "and." RealityCharting® provides a simple drop down menu that allows you to identify any causal relationship with "OR" logic.

Additional Aids to Communication – In large organizations a stakeholder commonly knows their role in the work process, yet no one knows the entire process. As a result, when we begin to solve a problem within our work process, we fail to communicate because our belief in common sense dictates that everyone knows what's going on. To overcome this problem, it is best to start the problem-solving process by creating a flow chart of the work process in question. Every player must be involved in creating the flow chart. A supervisor may create a strawman chart to get started, but every player should review and comment on the legitimacy of the flow chart. As the process diagram or flow chart is developed, errors, omissions, and misunderstandings will surface; and you can begin asking "why" and create a Realitychart. You may have to create or find pictures, diagrams, or drawings of equipment to help in the understanding.

Root Cause Analysis Defined

Taking everything we discussed in this chapter into consideration, we can define effective Root Cause Analysis as follows:

Root Cause Analysis:
Any structured process used to understand the causes of past events for the purpose of preventing recurrence.

An effective root cause analysis must:

1. Define the Problem
   a. Include the significance or consequence to the stakeholders.
2. Define the causal relationships that combined to cause the defined problem.
   a. Provide a graphical representation of the causal relationships.
   b. Define how the causes are interrelated.
   c. Provide evidence to support each cause.
3. Describe how the solutions will prevent recurrence of the defined problem.
4. Provide a report that clearly presents all of the above.

In addition to an effective root cause analysis, a complete corrective action tracking system should be in place to make sure solutions are implemented and validate their effectiveness. (See Chapter 5 for details.)

Solving Problems Effectively Using RealityCharting®

Using RealityCharting® software will produce a more accurate picture of your problems than ever before possible. For groups it creates a common reality and a visual map that facilitates dialog through appreciative understanding. By appreciatively understanding all perspectives, we understand the all important causal relationships that would not otherwise be found. Without the

Realitychart, we are left to linear thinking and storytelling. Therefore, we need to appreciatively understand all perspectives and use the tools described in this chapter to the best of our ability.

This is not always easy without some practice because appreciative understanding and Apollo causal analysis are a new way of thinking to most people. To overcome the anxiety associated with implementing this new way of thinking, RealityCharting® software will not only provide a step-by-step map, it will ensure an accurate chart every time. Those who use these tools report that it has fundamentally changed their lives by improving their day-to-day problem-solving skills.

The more we use these tools, the more natural they become and the long-term payback is a more accurate way of thinking because we realize things don't just happen. We see that every effect has causes, and most of them can be known and easily documented using RealityCharting®. The common reality resulting from this process allows effective communications between all stakeholders. No arguments and no confrontations.

Since problem solving is about effective solutions, the ability to get all stakeholders to buy into them makes the Apollo process unique in the world of problem solving. To learn how to ensure effective solutions, read the next chapter.

4

Identifying Effective Solutions

For every complex problem there is a solution that is
simple, neat, and wrong.
— H. L. Mencken

*Identifying effective solutions is the primary purpose of problem solving
and there are some subtleties we need to understand. Solutions act on
one or more causes in the cause and effect chain without regard for
position. So the notion of a "root" cause or a magic bullet at the end
of a chain becomes meaningless. If we must retain the notion of "root
causes," then they are the ones to which we attach solutions. And, since
any given problem has an infinite number of causes, there are an infinite
number of possible solutions.*

*What we need to find are the best solutions. As we will see in this chapter,
these solutions must meet the following criteria.*

1. *Prevent recurrence.*
2. *Be within your control.*
3. *Meet your goals and objectives.*

We have been led to believe that effective problem solving can be had by finding the root cause at the end of a chain of causes. On the surface, this seems to make sense; but on examination it is overly simplistic because it ignores the infinite set of causes discussed in Chapter 2.

As we saw in Chapter 2, our world is not linear and therefore this logic is overly simplistic and grossly ineffective. As we have seen, our world is made up of an infinite set of causes all connected through causal relationships. Some of these relationships are complicated by feedback loops. The beginning and end of causation are determined by our knowledge and understanding of the problem at hand.

Once we understand the non-linearity of our universe, limiting oneself to a linear understanding makes for terribly ineffective solutions. By understanding that there are an infinite number of causes connected in many ways, we begin to see that an infinite number of possible solutions exist. We may only need to affect one cause in a chain such that the problem does not occur or we may need to attack several.

These causes do not have to be at the end of a chain as the notion of root cause would have us believe. Indeed, the best solution may be to remove the very first cause, or as we have learned to call it, the primary effect. For example, for some people the best solution to human error would be to eliminate all employees. If this is possible and allows us to meet our goals and objectives, then it may be the best solution. But, we can't know that until we understand the causal set that governs the situation or system. The system may require human interaction, in which case removing all humans would not meet our goals and objectives. From this discussion, we can see that a solution can be defined as follows.

> **Solution:** An action taken upon a cause to affect a desired condition.

Generally, the action is to remove, change, or otherwise control a cause. Sometimes the action is to not act, such as not smoking to prevent cancer. However, this is merely semantics because the cancer is caused by smoking and the solution is to remove "smoking"; thus, we are acting on a cause.

The desired condition or outcome becomes the focal point of any solution. The purpose of problem solving is to establish conditional causes in the form of a solution or solutions that allow us to accomplish our goals and objectives.

For problems that have already occurred, a primary goal is to prevent recurrence. For problems that may happen in the future, the goal is to prevent occurrence. If our goals and objectives are to produce something safely and efficiently, then we need to know that before we set up the conditions that will allow that to happen. We must establish conditions such that when human actions occur, they will cause a safe condition and an efficient process or system. No system involving humans is perfect. As unacceptable effects occur in the process, we can go back and understand the problem causes and change them to accomplish our goals. In complicated systems where humans are involved, the number of variables is enormous; so, we are continually solving problems. This process varies significantly from problem to problem and group to group depending on the personalities and experience, so there is no right way to find solutions.

To understand the solutions phase of the problem-solving process in more detail, let's examine how to find solutions and then explore some guidelines for dealing with outliers.

Problem-Solving Process

To put things in perspective, let's review the Apollo problem-solving process. The four phases are listed below. The first two phases were discussed in Chapter 3. The focus of this chapter is on Phase 3. Phase 4 is discussed in the next chapter.
1. Define the problem.
2. Create a Realitychart.
3. Identify effective solutions. Challenge the causes and offer solutions. The best solutions must:
    Prevent recurrence.
    Be within your control.
    Meet your goals and objectives.
4. Implement the best solutions.

The basics of Phase 3 are as follows:

■ *Start on the right side of the Realitychart and begin challenging the causes. We challenge the causes by asking why is the cause here? What could we do to remove, change, or control it such that the primary effect does not occur? What does it mean? Is there more to this elemental causal set? Offer possible solutions for each cause and take note of them. RealityCharting® provides an easy way to do this. When challenging the causes, there are no rules; but there are guidelines that will make this step easier and more effective. Working top to bottom and then right to left is one of those guidelines. When finished with the upper right cause, move down to the cause below it in the same vertical position. When you get to the bottom of that column, go back to the top and move left one space, offering solutions for each cause along the way. The RealityCharting® Wizard Step 3, Identify Solutions does this automatically for you. If you want to address a specific cause on the chart, which often happens late in the investigation, a Solution Tool is provided that allows you to select any given cause and attach solutions to it at any time.*

■ *Make no judgments about possible solutions at this time. Move as quickly as possible from top to bottom and right to left. Do not waste time trying to analyze every solution at this time. Staying lighthearted at this stage is very helpful.*

■ *Work your way left to the primary effect challenging causes as you go. Be careful to look at every cause or elemental causal set. If nothing comes to mind, move on. Don't dwell on your inability to offer a solution for every cause. Sometimes a short linear causal set within the chart becomes the target for optimum solutions because the other cause chains are clearly outside your control or so incomplete as to require a major research effort to find all the causes. If this is the case, focus on that causal chain and scrutinize it thoroughly. The solutions phase often results in adding more causes to the chart because you realize you stopped too soon or missed some branches. Take the time to add these new insights.*

- *Be open to creative ideas. More on this later.*
- *After you have exhausted your creative juices and challenged each cause, check your solutions against the solution criteria. Again, RealityCharting® provides an easy way to do this. Once you have identified which solutions meet the criteria you can select which ones you want to implement and they are automatically placed in a report.*

Now, let's examine in more detail the process of identifying the best solutions.

Solution Criteria

Everyone has their own opinion and favorite solution, as discussed in Chapter 1. What makes one solution better than the next one? Or, a better question is what makes one solution more effective than the next one?

No matter what the event-based problem is, the solutions must have certain characteristics for them to work, and "work" is the operative word here. In a training exercise we ask students to evaluate a problem and identify what they think is the best solution. This is always an interesting exercise. At this point in the training, they are still operating out of the silly notion that there is one right answer. And, of course, the right answer is usually their answer. They argue with one another and use various forms of persuasion to get their solutions accepted by the group. They fail to recognize that there is no such thing as a right or wrong answer to event-based problems. When asked what characterizes an effective solution, "the solution must work" is the most common answer. But what does "must work" mean?

After analyzing over 25,000 answers to this question, I found the answers will always fit into what we can call solution criteria. The best solutions must meet the following three criteria:

1. Prevent recurrence; to include similar occurrences at different locations

2. Be within your control
3. Meet your goals and objectives, where goals and objectives should include:
   a. A solution that does not cause other unacceptable problems.
   b. A solution that provides reasonable value for its cost.

Assuming we are dealing with a problem that has already occurred, preventing recurrence is essential. Preventing recurrence means that it does not happen again for the same (known) set of causes. Anything other than this is a failure to understand the problem. While this is not always possible, we should strive for a 100% non-repeat.

Solutions must be within our control or they will not work. It is a common human tendency to identify solutions that require other people to act. If you are using the Apollo process correctly, this will not happen because all stakeholders will be involved and everyone will take responsibility for the solutions within their control.

Sometimes a solution does not appear to be within your control because it requires higher-level approval. If your cause and effect chart has been prepared properly, you will be more likely to "sell" the proposed solution to those who have the authority. The Realitychart can significantly expand your sphere of influence because it is not just another opinion-laced story. It is evidence-based causal relationships, which are hard to ignore.

The criteria that causes such a wide variety of possible solutions is the need for solutions to meet our goals and objectives. Most businesses exist to make money, so the solutions should provide a maximum return on investment (ROI). Most companies have an established ROI requirement before implementing a solution. An alternative to measuring ROI is to measure the number of problems above some threshold criteria, for instance, all events costing more than $50,000 in lost revenue. If over time the number of events exceeding this level of concern goes to zero, then your problem-solving process has been effective. If your company is interested in continuous improvement, the threshold criteria should be evaluated and changed according to your goals and objectives.

An important aspect of the solution criteria is the "you" or "your." Your control and your goals mean the solution is owned by those who are going to be responsible for the failure, no one else. The implications of this are multifaceted. It means that no outside organizations have the right to second-guess the solutions unless they are willing to accept the consequences of failure. It also means all stakeholders must understand what their goals and objectives are before they can be expected to be effective problem solvers. Many employees simply have no idea what their goals and objectives are, so they have been set up to fail as effective problem solvers.

Every solution is directly related to the purpose for solving the problem. When employees interject their own purposes into the solution, they may or may not coincide with the purpose of the team or organization. Look for these biases when evaluating solutions. For example, if the purpose of an organization is to operate safely, the solution to a production problem cannot include creating a safety hazard. Unfortunately, we humans find it difficult to recognize all our goals and objectives when considering problem solutions. We tend to be myopic and only recognize one goal or purpose for a given problem. While many organizations today have a set of company goals in the form of mission statements and strategies, these are rarely internalized by every employee.

Each member of an organization understands their role differently. In large organizations employees often see themselves as part of a group. Engineers may form their own understanding of what their contributions to the company goals are, and it is different from the other groups. Operators, maintenance people, and sales people all see their contributions differently. They form a group identity that sets them apart from other groups. When it comes time to work together to solve company problems, people may be divided into different camps looking out for the interests of their group. If you find this thinking, it is essential that you write down the goals and objectives associated with the problem at hand. One obvious goal is to prevent recurrence, but other goals may be to make a 20% ROI or provide the best service possible. Each company goal needs to be defined in more detail relative to the problem you are working on.

Because goals and solutions are so closely related, finding the best solutions often becomes an iterative process of discovery. The causal relationships begin to fill out with more causes between the causes, and a greater clarity of the problem as potential solutions are discovered.

Multiple Solutions

If you followed all the rules in constructing your Realitychart, preventing recurrence will be unquestionable regardless of which solution(s) you chose. However, in some cases 100% assurance is not guaranteed because your ability to control the cause may be limited, such as only being able to slow a leak rather than eliminate it. In these cases one solution may prevent recurrence 90% of the time, while another solution will prevent recurrence another 9% of the time, thus giving you an assurance of 99% prevention. If this meets your goals and objectives, that may be where you stop. Only you can set these standards of quality and excellence. Depending on the significance of the problem, you may be happy with 85% assurance of a non-repeat.

Solution Guidelines

In the course of identifying solutions, often some solutions do not meet all three criteria but still provide value. We may choose to implement them and this is acceptable, but make sure to identify which cause it attacks. Also, identify these solutions for what they are. If they are not required to prevent recurrence, but will improve the situation, then make this clear. Sometimes we include pet peeve solutions into an inappropriate problem, which results in spending money on something that does not prevent recurrence.

It is not unusual to discover a very good solution that does not seem to be connected to any of the causes on the chart. The solution seems to come out of nowhere. (*Nowhere* is defined as the subconscious mind that is always at work on our problems.) When this occurs, it helps to ask what causes does the solution attack and follow this line of questions to discover a whole new set of causes. Usually we are able to connect the solution with our existing Realitychart, but it may take some exploring. We often know things in a visceral sense, called gut feel or intuition. This understanding of the world is held in the part of the brain known as the emotional center or limbic system. It is not well connected to the cerebral cortex and the language center, so we don't have a direct command of it; but it is there nonetheless.[1] Do not ignore these feelings because they may hold important insights to effective solutions. Try to document them on the Realitychart.

Solutions should always be specific actions. It makes no sense to attack a specific problem with nonspecific solutions. If the problem was "no money," the solution shouldn't be "get some." Do not include solutions such as review, analyze, or investigate. Such solutions are a copout as they are really saying we don't know what the problem is and won't know until we can gather more causes and evidence. Avoid this denial and state that you don't know what the causes are. Implement mitigating solutions until you can investigate further.

If an ancillary solution is to further investigate or review and the main solutions will prevent recurrence, then this is acceptable, but list these ancillary solutions on a separate tracking list. This will avoid the corporate auditors' complaint that you have not completed this commitment.

Avoid solutions that include the prefix of "re-," such as re-train. Avoid the favorite solutions such as the following:

- *punish*
- *reprimand*
- *replace the broken part*
- *investigate*
- *revise the procedure*
- *write a new procedure*
- *change the management program, that is, re-engineering*
- *redesign it*
- *put up a warning sign*
- *ignore it—stuff happens.*

This list contains the most common favorite solution categories I have found over the years. It does not mean an effective solution is impossible if it comes from one of these categories. What it does say is that you are in a rut and chances are the problem will repeat itself. Favorite solutions usually mean you have also identified your favorite set of causes. Go back and look closer at the Realitychart. Look for branches, conditional causes, and causes between the causes. Chances are you missed some key branches or assumed a causal relationship that is not well understood.

Sometimes the solution may be to do nothing, for example, if the Realitychart reveals that the causes are unique and the probability of repeating is low. It may be that the consequences as identified in the significance portion of the problem definition are minimal. Again, always consider the solution criteria as a function of your purpose. The right answer is the one you choose as long as you can honestly say it meets the three criteria.

The unthinkable may also happen. My studies show that in industry about 5% or 6% of the time we are not capable of finding a solution to our daily problems. This number seems to be consistent across different industries, and I believe the reason is that our knowledge of the processes we control is limited to about 95%. These statistics reflect the fact that sometimes we simply don't know what happened. This is not to say we won't know in time, but we stop looking because the task is too expensive, time consuming, or difficult. When we find ourselves in this condition, we should devise a plan to capture more information and causal evidence. If the problem happens again, we will know more causes.

Solution Killers

"A pessimist sees the difficulty in every opportunity; an optimist sees the opportunity in every difficulty."
— Sir Winston Churchill

Some people are simply not happy unless they are complaining about something. These chronic complainers see the world as one big problem, always complaining or putting others down. They are the only ones with the right answers and if the world would only stop long enough to ask their advice, everything would be wonderful. I suspect you work with some people like this; and if you do, you need to know how to overcome this negative attitude so the rest of the organization can accomplish effective solutions. Here is a short list of some common expressions used by these naysayers:

- *"It will never work here."*
- *"We're too busy for that."*
- *"No one will buy it."*
- *"We already tried that once."*
- *"That's not our policy here."*
- *"It isn't in the budget."*
- *"Good thought but impractical."*
- *"Top management will never go for it."*
- *"No one else is doing it that way."*
- *"Wrong!"*
- *"We've always done it that way."*
- *"Good idea, I'll get back with you"* (and never does).

Solution-killer statements are usually caused by the fear of change—the result of the natural process of groovenation. Some people are more groovenated than others. Groovenation starts in the teenage years and all adults have it to varying degrees. Notice that children don't have this affliction. In fact they have exactly the opposite condition. They seek the unknown and welcome change as a gateway to experience, even at the risk of endangering themselves. If we can get the fearful people to be more like children and set aside their fears for a few minutes, they might be able to see other possibilities.

Never let a solution-killer statement go unanswered, no matter who says it. This can be done tactfully by focusing on learning, growth, and improvement, not change. If the boss says a solution cannot be implemented because it is not in the budget, then redirect the focus to the purpose of doing business. Make sure you have done your homework and can show the ROI, the inherent value of the solution, or the requirement to fix the problem.

A common solution killer in business is related to money. I am amazed by how many middle-level managers do not understand the purpose of business is to make money, not save it. Saving money is for governments and individuals on a fixed income. The rest of us are trying to make money. The fundamental process of doing business requires that you spend money to make money. If your solution can be shown to make money for the company, whether it is in the budget becomes a moot point. Any bank will lend money to an established business if they can show that the ROI is adequate. Look for other alternatives such as scheduling implementation over a longer period of time or until the money can be put in the budget, but do not let this solution-killer go unchallenged.

The purpose of the killer phrase is to stop discussion, and it works quite well if left unchallenged. If you are interested in effective solutions, it is imperative that you speak up. The best response is to ask the person making the statement to "say more about that." Pause, and if you don't get a response, ask another question like, "What do you mean we tried it once? Tell me what you did last time? Or, why did it fail?" Often, these people have no idea why the solution you are suggesting would fail. They will have opinions, but they will not have a cause and effect chart that documents the cause and effect relationships.

It often helps to play dumb; be the student and ask them to teach you, to explain why the solution will not work. As they provide answers, turn those answers into causes and see how they fit into the Realitychart. If they have valid concerns, they will fit into the common reality you are creating. In the process of adding their causes to the chart, you will have gained an ally. Your purpose in this strategy is not to prove the naysayers wrong. Your purpose is to get them to forget their fears long enough to play the game—to engage them in cause and effect thinking. By engaging them, you help them lose their fears and become confident with the new understanding they helped create. Once they see how well the process works, they become more open to it and will be more engaging thereafter.

Never get into an argument about who is right or who is wrong. There is no such thing in the world of events; the solutions are only good, better, or best. We never have enough information to totally understand anything we do; we can only operate on what we know. If it is documented on the Realitychart, at least we have a form of communication that allows for everyone's understanding to be represented. Obtaining this common reality is the key to effective problem solving, communication, and buy-in from all stakeholders.

Creative Solutions

Creativity and logic have always appeared at odds. They shouldn't be, but it is rare that a person is both very logical and creative. Like a flower, creativity can be nurtured and bloom or be nipped in the bud by the caregiver. The creative process is one of absurd connections, whereas the reasoning process is one of structure and rightness. Reasoning and creating seem to be on opposite ends of the mental spectrum, and indeed they cannot occur at the some point in time. This is not to say we can't jump from an analytic state of mind to a creative state in short order because we can, but the mere act of making absurd connections while laying out a clear set of logical connections may be impossible.

Regardless of the solution, creative or logical, we must understand the causal relationship between the solution and the primary effect; or we are back to guessing and voting, which have a low success rate. How can we go from one mental state to the other? Let's explore some strategies I have found to be very helpful.

Listen for the Laughter

Laughter is caused by the improbable connection of two or more things. "Of all the things I lost over the years, I think I miss my mind the most." If you find this funny, it is because you would never have thought of the possibility of losing your mind in the physical sense because it is secured between an impressive bone structure. It simply can't be lost. Yet we use "lost his mind" to describe people when they act strange. Since laughter and creativity both consist of heretofore unmade connections, it stands to reason that spontaneous laughter may lead us to creative solutions.

The next time you hear laughter while discussing solutions, stop and find out what caused the laughter. It usually appears to be something that is so absurd that you won't bother to take it any further. Do it anyway. Ask the person who caused the laughter to explain why doing whatever it was they suggested would have any effect on the problem. Don't ask them to explain why they thought of it and be very careful not to make any judgments. They may or may not be able to tell you why they said what they did, but look for a cause that is being removed, such as "fire the boss, yuk, yuk." You might ask what would firing the boss do? "Well, it would allow us to do our jobs right." "Does this mean we aren't doing our jobs right?" "How could we do them better?" "Are there barriers in the way?" As you go down this uncharted path, suspend judgment in a positive sense. That is, look at all statements as eventually leading to something positive, even if they initially appear not to.

Rapid Response Method

Another way to find creative solutions operates on the premise that since we cannot reason and create at the same time, we need to find a way to turn off the reasoning and only allow the absurd connections to come into our mind. Regardless of how creative people are, logic and reasoning usually creep back into the thought process. This is true except for the very creative, who have the ability to suspend all judgment and just go with their feelings. One way to force this activity is to not allow the mind time to analyze what is being presented. This requires a group of people and a facilitator who rapidly extracts ideas from the group, starting with the completed Realitychart. Use the following steps to find creative solutions.

Step 1: Create possible solutions.

Some call this step brainstorming. It is in a way. However, it is different in that we are addressing a specific cause from the Realitychart one cause at a time. To accomplish this, we need to establish some simple rules.

Rule #1: Speak only when it is your turn. The facilitator will let you know when it is your turn. No comments are allowed by anyone else for any reason during the creation session. If you have nothing to offer, you say "pass."

Rule #2: Do not explain your ideas. Again, that is an act of rationalizing. Put it on a bumper sticker and move on, quickly.

When I say work fast, I do mean fast. It may take two people writing on flip charts (or cards or self-adhesive notes) because one cannot keep up and you don't want to slow down the group. If they slow down, they start thinking. You go around the group for the cause you want to challenge until all participants have said "pass." Then move on to the next cause.

Step 2: Evaluate the effectiveness of the proposed solutions.

Go back and look at all the proposed solutions. Here, each person can explain their ideas as much as they want. Anyone can help them look for the connection if they have difficulty expressing it. When they hit on something interesting, the group begins to synergize. At this point, appreciative understanding and other creativity tools can be employed. It is not unusual to discover your Realitychart, which was perfect a few minutes ago, is woefully inadequate at this point. Revise it as necessary.

If the group is tired from performing a causal analysis, it may be best to have the rapid response solution session some other time. Let problem significance guide you with respect to timing, how hard you work the group, and so forth.

The Gano Rule

Creativity often requires the use of what psychologists call the unconscious mind. Using the unconscious mind is best accomplished by sleep and play. Unless time is critical, take as much time as you have to find the best solutions. After creating the Realitychart, sleep on it before you start the solutions phase. I have a rule that I always follow when making important decisions: I always sleep on it, unless I am forced to do otherwise. I call this the Gano Rule and salesmen don't like it. If you spend time trying different solutions and can't find any that really cause you to go wow!, sleep on it.

If you can't sleep, go play. Our unconscious mind is busy working on our problems while we play. Everyone has experienced the great aha's! that seem to come out of nowhere. The morning shower is a great place because the brain is usually not yet connected for the day; and we heat it up with hot water and the blood flows better, causing ideas to race through our minds. Sometimes those ideas seem strange, but they are simply new connections of thought. Evaluate after you get out of the shower, enjoy the moment of creativity if it's there, and write it down or try to explain it to your mate. This will help solidify it in the "real world."

Behavioral studies have shown that about 95% of all five-year-olds are highly creative, according to Charlie Palmgren of SynerChange International.[2] Then, something happens between the ages of five and eight. Less than 5% of those older than eight are highly creative. The research seems to indicate that formal education may be contributing to this lack of creativity. When children enter school, their learning goes from that of natural play to a structured, scheduled, and homogenized process based on the false premise that everyone is the same.

Failing to understand that play is the natural learning process, we force-feed children meaningless facts and rule-based schemes. If left to their own strategies, the child will solve problems by alternating between working on the problem and playing at something else. While the child is playing, the unconscious mind is working on the problem. When the child returns to the problem, the solution often just appears out of nowhere. As adults, you have probably experienced the same thing. Play is important to the natural problem-solving process, so we need to incorporate it into our problem-solving plan.

Yes-Anding

"Yes-anding" is a creative strategy that helps us avoid negative judgments throughout the solutions stage. It works like this. Avoid using the word "but." When a fellow team member says something you disagree with, try to empathize and then agree with it. Follow your profound agreement with the statement, "Yes, and we could add to that with [insert your idea.]" "Yes-anding" can be used to brainstorm by asking every person to build on the expressed solution. It moves very quickly, like the Rapid Response Method, so we do not want to slow down the process with logic or analysis at this point. Be enthusiastic about the previous solution even if it is totally off the wall, such as: "Oh, yes!" "Wesley, that is the most incredible solution I have ever heard AND we could build a miniature mousetrap from your new tennis racket by adding a spring at the bottom."

The "yes-and" strategy is based on what children do when they play. Young children have not yet learned to put each other down. They have no experience on which to judge, so they accept what is stated as being really cool and then build on the thought until they have turned a cardboard box into a castle complete with kitchen and cannons.

Creative Solutions Example

The following real-world problem taken from a U.S. process plant provides an example of the creative process using the Realitychart.

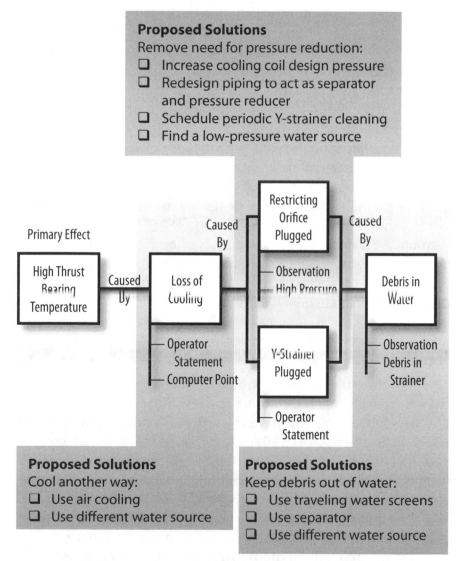

Figure 4.1. Creative Solutions

The problem was a high thrust-bearing temperature on an 800-horsepower river-water pump motor. This is a vertical pump and motor at a remote location, and the original investigation found river debris restricting an orifice in the cooling water line to the oil reservoir. The bearing cooling water came from the pump discharge and often carried river debris. To "solve" this problem, the engineering staff implemented their favorite solution for plugged water lines. They had a Y-strainer installed. With this solution, each time a high-temperature alarm occurred, an operator traveled several miles to the pumphouse to clean the strainer.

A few years later, a reliability engineer formed a new team to investigate the problem using the Apollo problem-solving methods. After creating a simple cause and effect chart, the team challenged each cause and offered several solutions for each. Their solutions are shown in Figure 4.1.

After evaluating the various proposed solutions, the one that best met the three criteria was to use a different source of cooling water, which they found about three feet away. The new source of cooling water was not only clean, it was also at low pressure, which allowed them to remove both the restricting orifice and the Y-strainer. Unlike the favorite solution of adding a Y-strainer, this did not add to the maintenance and operational workload. It also improved the reliability of the motor by removing equipment and providing a cleaner cooling water source.

Identifying Effective Solutions

Effective solutions mean we must prevent the defined problem from happening again. If you are experiencing repeat events in your life or organization, you have missed the fundamental tenet of effective problem solving. Effective solutions are ones that work for you or your organization, not someone else. A solution will work if it meets the three criteria of preventing recurrence, is within your control, and meets your goals and objectives.

The difference between conventional wisdom and the Apollo problem-solving process is the belief in a right solution. As we have learned so far in this book, there is no such thing as one right answer to event-based problems; only good, better, and best. The best solutions not only meet the solution criteria, but they are the ones you or your organization choose. By involving key stakeholders in the problem-solving process, you will create a common reality from which effective solutions become obvious to all. If anyone else involved in the decision-making process has questions or doesn't understand the causes, they should be invited to add their reality to the chart so everyone can gain from their different perspective. When this happens, the solutions are more likely to be better. Removing the conflict found in most team-based problem-solving endeavors means effective solutions are more likely to get implemented because the stakeholders have ownership.

While the solutions are often more effective than other problem-solving processes, we need to be aware that a few people will not be willing to change their belief systems. As Henry Ford once said, "Whether you think you can or think you can't, you are right." By seeking to engage all perspectives and following the basic rules of the Apollo process, we can usually bring these naysayers into the fold and get effective solutions agreed to by all players.

Remember to listen for the laughter when considering solutions. There is always a smart aleck in every group, so listen for the telltale signs of laughter. Follow the reasons behind the laughter and look for the cause of the absurd connection. It is the creative solutions that are often the best solutions because they have identified a cause path that has never been understood before. As you become more and more proficient in using the Apollo process, you will experience some interesting interrelationships between problem definition, causes, significance, and solutions. They all seem to work off each other and since it is so easy to share the common reality and make changes with RealityCharting®, better solutions are more abundant.

Now that we have learned how to define the problem, create a Realitychart, and identify effective solutions, we can explore the bigger picture of an effective problem-solving program. The next chapter will provide details of such a program and discuss the fourth phase of the Apollo process: implement the best solutions.

References

1. Carter, Rita 1999 *Mapping The Mind*, University of California Press, London, UK
2. Obtained from various presentations by Charlie Palmgren, SynerChange International, Atlanta.

5

Implementing Solutions

"It's a funny thing about life: if you refuse to accept
anything but the very best, you very often get it."
— Somerset Maugham

*The last phase of the Apollo problem-solving method—implementing
the solutions—is as simple as flossing your teeth. But, like flossing, it
won't happen unless we assign value to the task and take action. What
we need is a program that shows inherent value simply by using it. In
this way, all stakeholders will see the value and want to implement
it. Effective problem solving then becomes institutionalized as a way
of thinking, not as an appurtenance or new program of the month.
This chapter lays out the basic elements of an effective program
centered around the Apollo problem solving method. The key elements
of an effective problem-solving program are as follows:*

1. *Dedication to Continuous Improvement*
2. *Comprehensive Problem-Solving Training.*
3. *Threshold Criteria—When to Perform an Analysis.*
4. *Simple Reporting Scheme.*
5. *Corrective Action Tracking Program.*

*With an effective problem-solving program, tracking and trending root
causes eventually becomes a thing of the past.*

If you can, think back to when you were in the first or second grade. Somewhere in those years you probably had an experience that would change your life forever. Your feelings were hurt in a way you had never experienced before. Later, you would learn this experience is called embarrassment. It usually happens when someone—a classmate, a teacher, or a parent—makes you feel stupid. You asked what seemed to be a legitimate question, and the next thing you know, you are ridiculed.

Eventually, you developed a strategy of being more careful when asking questions. Perhaps you learned to test the water with simple questions, just to see if you got hammered. If you did, you probably laughed and backed off. As you developed various defenses to protect your feelings, you may have developed a go-along-to-get-along attitude because being liked is important. After all, being liked fills the greatest human need of all: the need to be needed. With this successful coping strategy, you "fit in" and became a good student, a good citizen, and a good follower.

Later in life you got a job and became part of the work force. You were eager and ready to become a valuable contributor to the company or organization. You were enthusiastic, asked lots of questions, and learned from your fellow employees. After about six months or a year you began to realize that as a new employee you could only ask "dumb questions" for so long. If you went too far with your questions, you quickly learned it could cause that same old feeling of embarrassment. You may have learned that nobody really cared what you thought because everything was already figured out by "smarter people," and again you became a good worker, a good citizen, and a good follower; you "fit in."

Your world was well defined and everything was going just fine when along came downsizing and reorganization. Suddenly, getting along was not enough. You were asked to challenge what is and look for new and creative ways to replace the status quo. You were told to do more with less; and you did what you knew, which included working harder and longer. New ideas with strange buzzwords like Root Cause Analysis and Socialization just kept coming. You may have stumbled, or you may have fallen out the door. At the very least you were confused by these new demands. You were unprepared because you stopped learning and subsequently may have forgotten how.

Whether this scenario describes you or not, we have all been set up to fail by the many years of conditioning that started back in the first or second grade when embarrassment first killed learning. Our culture discourages a questioning attitude, and adult learning is too often viewed as having little or no value. Conventional wisdom holds that we can just learn as we go along, and this allows us to get along. If good enough is your goal, then this strategy will work, but it will severely limit your effectiveness.

We have failed to understand that this brave new world of high-performance work teams and socialization requires at its foundation an absolute unwavering dedication to learning. Learning requires an open-minded environment where questions are encouraged without any embarrassment, and where answers can be judged on their ability to add value to our lives—to help us meet our goals and objectives.

If you want to know what learning is and how exciting it can be, please visit a preschool tomorrow. Just ask to sit and watch. These youngsters don't yet know what embarrassment is because their minds haven't been polluted with the destructive duality of good and bad questions.

The Apollo method provides an excellent tool to enable a learning culture, but it first takes a commitment from everyone to embrace learning. Learning to be better problem solvers requires an unwavering dedication to continuous improvement and comprehensive training in basic problem-solving skills. More so today than in years past, many companies are recognizing that good problem solving skills like the Apollo method are a core competency.

Commitment to Continuous Improvement

To exist is to understand the here and now; to grow and prosper requires a commitment to learning. While a fundamental tenet of the quality movement since W. Edwards Deming, the renowned quality management consultant, continuous improvement is much harder to do than to say. The human mind is designed to establish patterns that cause success, and once we find a success path it is hard to change. We simply do not like change of any kind. It is only the few who actually experience the value of change who understand this concept.

The willingness to change is a lesser force than the need to maintain status quo. I saw this in action when I helped conduct a review of a corporate safety audit at one of our client's plants. We were trying to find the causes of a significant breakdown in the safety program. Along the way we found many causes, but as the review continued, I began to notice increasing evidence of a failure to learn from mistakes. In fact, I noticed a total lack of respect for learning. This facility had a proud history of being Number One in the Corporation when it came to safety. Having succeeded, they came to believe they had a formula for success and could ride on that formula forever.

As environmental laws changed and new safety requirements were implemented, this organization wrote procedure after procedure to accommodate the new requirements, but little actually changed. When asked how they knew that laws and standards were being implemented, every department head said because it was in the Safety and Health Manual. After I repeated the questions a few more times, they all admitted they had no idea how they met all the requirements set forth in the manual. Indeed, many of the requirements were not being met. An internal audit had revealed their noncompliance and it was ignored for almost a year until their corporate audit team arrived onsite. Groovenation rules!

Learning requires a questioning attitude. Without a questioning attitude we fall into the human trap of groovenation, where our minds seek the familiar and reject anything new or different than our existing paradigms. Another investigator and I witnessed this as we were leaving the plant following the review. On the way out we passed two fire hydrants near the cafeteria. Both were rusted, but one had a severely bent box wrench hanging on the side. This wrench opens the valve to supply fire water when needed. We stopped to ponder the implications in light of the Realitychart we had just finished. The hydrant was one more piece of evidence of the causes we had found, but more fundamentally, it exemplified the lack of a questioning attitude. We wondered out loud how many other conditional causes lay in wait of an action to cause significant consequences. As the workers walked by this condition each day, on their way to the cafeteria, they saw nothing out of the ordinary.

We looked at it with fresh eyes and asked what caused a forged wrench to be so severely bent? Could it be a sticky valve that may not open when needed? Whatever the cause, it is clearly beyond the design basis of the wrench, yet it goes unnoticed. Why?

What I found, which is common to most companies and industries today, is a corporate policy that advocates a commitment to continuous learning yet punishes or derides anyone who speaks up about potential problems. What we must ask is why these seemingly intelligent people do this. Certainly, the cause has to do with group dynamics and the notion of a collective consciousness whereby we suspend our individuality for the perceived benefit of the group, but I think it is more than this. In organizations where effective leadership exists, they do not have this problem, so I believe a deeper cause is the inability of the leadership to deal with questioning attitudes.

Effective leaders understand the value of a questioning attitude and incorporate strategies to encourage it. However, it has been my experience that effective leaders do not grow on trees and most managers and so-called leaders simply do not possess the skills to resolve different perspectives. Quite often, the management strategy is "my way or the highway."

The Realitychart can be used to change this inability to deal with conflicting ideas and perspectives. By providing a simple tool that allows for all perspectives to be viewed in one picture, different realities can be discussed in an argument-free environment. By communicating with evidence-based causes, we eliminate ineffective storytelling and inference by categorization. We replace these old strategies with a simple tool that encourages a questioning attitude and differences of opinion.

Another example of the difficulty of continuous improvement was demonstrated to me personally when I sent a paper about the Apollo Root Cause Analysis process for publication in a quality assurance trade magazine. After months of review by the peer review committee, I was told this method was not root cause analysis because it did not include the popular Ishikawa Fishbone diagram, which is the foundation of problem solving in many quality assurance programs.

The reviewers had established a fixed view of the world and were not going to be mislead with anything new. Yet these same people are the ones espousing continuous improvement. This is not personal failure on the part of these individuals but rather part of a greater tragedy of the human condition caused by the natural tendency toward biased thought. The majority of us simply can't help ourselves. It seems to be fundamental to the nature of the organism.

To help your organization break out of this trap, a learning environment must be established. At the heart of this environment is a new way of communicating that uses the Realitychart as the basis for decisions; however, a new philosophy also must be established that is based on the need for improvement. By focusing on improvement, not change, people are more likely to accept change.

Educating employees about the cause and effect principle will help them learn that things do not just happen. Everything has a cause and only our ignorance prevents us from knowing the causes. By knowing that every effect has conditional and action causes, they begin to see all the conditional causes around them and wonder what action cause will come along and cause an undesirable effect. Or, what action they can take that will combine with the conditional causes they observe to cause the desirable effect they seek.

By understanding that there is an infinite set of causes, again limited only by our own ignorance, we can begin to overcome the arrogance associated with right-minded thinking that has been shoved at us all our lives by the various institutions that we established to "educate" us.

With a firm understanding of the cause and effect principle, a philosophy that values everyone's perspective is possible. From this philosophy comes the understanding that continuous improvement is not only possible but is the best course of action. We can let go of our fear of change and seek the unknown with the full knowledge that the infinite nature of the cause and effect principle will allow and encourage effective problem solving.

Institutionalizing the Process

The Key is RealityCharting® & Training

As an engineer, I was called upon to solve many technical problems. I can remember my fear each time I was called. I was well trained to know the laws of physics and how they applied to daily life and the industry in which I worked, yet I was always anxious. My anxiety was always based on the fear I would not have the right specific knowledge to solve the problem.

After internalizing the cause and effect principle and recognizing the infinite possibilities for solving any problem, the anxiety is gone and I now attack all problems without fear of failure. While I know I don't know what the causes or solutions will be, I do know that I can take the expertise of those who work on the problem and create a Realitychart to find the best solutions every time. With each success and no failures, the need to cling to a fixed set of rules and prejudices, falls away. A sense of freedom replaces the fear of failure. The diversity of others' thoughts becomes the pieces of a common reality that enables learning.

Institutionalization of effective problem solving is best obtained by providing a comprehensive training program to employees along with an enterprise version of RealityCharting® software. The Enterprise version is a browser based application that can be customized to interface with your current reports and tracking system. Follow these steps:

1. Train all managers in the fundamental principles and the basic tools. When managers understand the Apollo problem solving process, they immediately see the value added and demand it be used for all problems.

2. Train everyone who solves problems on a daily basis to facilitate problem solving. This includes all supervisors and lead persons. When trained, these people become the program champions. They become very efficient at problem solving and gain respect from all levels in the organization because they continue to show results never seen before.

3. Train the general employee in the Apollo problem solving process. Anyone who might be involved in problem solving should know about the cause and effect principles and the basic tools of preparing a Realitychart. This training teaches the general employee that storytelling is generally a waste of time, that everything has a cause, and that those causes come in the form of conditions and actions. They learn to create Realitycharts and to separate opinion from fact. In time they learn to think causally.

4. Use RealityCharting® to guide you through the process. RealityCharting® ensures that an accurate Realitychart is created and the Apollo problem solving method is followed every time. Because of the high quality charts, managers and supervisors spend less time reviewing process and structural issues and more time on the content of the analysis. RealityCharting® users become more proficient problem analysts and because the findings can be easily shared, a more diverse set of stakeholders can be involved. RealityCharting® facilitates the creation of a common reality resulting in more effective solutions. Build your best understanding of the problem and its causes, e-mail it to others and interact to develop a more inclusive analysis.

At facilities where this complete training has been implemented, employees often work informally on daily problems by finding a vertical surface to create their common reality, based on input from all stakeholders. As they reach their point of ignorance on each branch of causes, they go back to the workplace and incorporate the natural human learning process of observing causal relationships until they can fill in the blanks on their chart. RealityCharting® is used to document and share findings with all stakeholders. As time goes by, they eventually have a very good representation of the causes of their problem and can now effect changes to prevent recurrence. The tools are simple enough to be used by anyone on any problem and at any location.

Prior to spending money, most companies want evidence that their training dollars will provide a significant ROI. Each problem is different; and while I can recite examples that have earned companies millions of dollars on one problem, the average ROI for each student that completes a facilitation training class is about 3,500% or approximately $25,000 within the first month following a class. This figure is for manufacturing and process industries and does not account for repeated successes, which directly multiply this ROI. Similar successes occur in safety-, quality-, and service-related problems, but it is harder to assess actual dollars earned.

The ROI for training is always hard to measure. If the training provides specific work skills that directly affect the ability to perform a task, there is no question about the cost of training. It is a go or no-go situation, and no cost assessment is required. But with training such as learning problem-solving skills, the advantages and payback are not always so clear.

Among the many cases of payback, one I have tracked for several years comes to mind. Using the Apollo methods for the first time after taking an Apollo training class, a reliability engineer was rewarded by fixing an old problem. He started his investigation with a piece of equipment that had failed thirty-five times in its lifetime. It had failed twelve times during the previous year. He assembled a team, gathered information, and prepared a Realitychart. Within a few days, the team had developed a new understanding of the problem from the common reality of the cause and effect chart. The solutions they implemented prevented recurrence.

When I spoke to him ten months after the corrective actions were implemented, there had not been a single failure. Three years later, they have still not had a failure. Since previous failures cost a minimum of $15,000 each, correcting this one problem has saved hundreds of thousands of dollars. He continues to use the Realitychart to drive his problem solving and continues to have successes like this one. Prior to his training in the Apollo methods, he had been using the conventional methods discussed in the Appendix. These methods simply don't work because they don't follow the cause and effect principle. For a comparison of common root causes analysis methods see the Appendix in this book.

When to Perform an Analysis

In every organization an incident investigation policy should be established to determine when a Realitychart should be created, and everyone should understand and buy into it. I have found three basic elements make it work: threshold criteria, evidence preservation policy, and clearly defined responsibilities.

Threshold Criteria

When we set about to establish a structured problem-solving process, the inevitable question arises: when should an investigation be performed? The simple answer is whenever you encounter unacceptable consequences. After adopting the Apollo problem solving methodology, the question becomes less important because trained employees will adopt it as a routine part of their job. The question then becomes when do we need to document the analysis process? Every organization needs to establish their own threshold criteria or sentinel events to answer this question. These criteria are a function of the industry and the organization. A service company will have threshold criteria such as "customer complaint" or "missed a goal." Manufacturers will have criteria like "total cost greater then $10,000" or "mean time between repair less than one year." Government organizations that enforce standards and laws have a ready set of criteria and only need to refer to any discrepancies or violations.

If an organization is properly using the Apollo method, there will be no more repeat events. Therefore, the number of events reaching the threshold criteria will eventually go to zero. Long before this happens, the criteria should be revised. In keeping with the spirit of continuous improvement, the threshold criteria should be periodically reviewed and revised to match the time available to perform investigations. If the time spent on investigation exceeds the time needed to operate the business, then the criteria is way too tight—relax it. If, on the other hand, no problems are meeting the threshold criteria, tighten them up. This periodic review must be part of an incident investigation program.

Evidence Preservation Policy

Establish a policy that requires preserving evidence. For example, broken equipment should be quarantined until experts can examine it and gather data. All too often in manufacturing and process plants, a piece of equipment breaks and because the organization is so efficient at fixing broken parts, a new one is installed and the old part discarded before anyone else knows of the failure. The same thing occurs in service industries when problems go unreported. Customer complaints are resolved, but no analysis is performed to find out why the problem occurred.

The efficiency at fixing broken parts is born out of the preventive maintenance or "broke-fix" mindset of the past. If we focus on reliability rather than repair, there will be no repair unless it provides the most cost-effective option. With this perspective, all failures are understood and measures are taken to prevent them, such as by replacing critical components before they fail or running noncritical components until they fail. This perspective also includes an Apollo root cause analysis on all unexpected failures, which requires preserving evidence.

Use an evidence preservation checklist (explained in Chapter 6) to obtain all relevant data, including personnel statements. Having a plan to gather as much information as possible immediately after the event is equally important. Identifying key people and giving them the responsibility to gather data as soon as possible can mean the difference between knowing the cause and the big shrug that occurs when the cause isn't known.

Responsibilities Defined

Establish committed investigation personnel to be on call twenty-four hours a day if your operation runs all day. They should start gathering data as soon as all safety issues have been dealt with. These people can be properly trained shift personnel or employees living nearby. A single individual should be given the responsibility to start gathering data and make sure all others are doing their assigned tasks. Everyone and anyone should be available to help if the event meets one of the threshold criteria. If there is no commitment here, then the threshold criteria may be too low. That is, if the problem does not warrant the time to fix it, then maybe it should not be worked on. Again, this is part of the questioning attitude needed for continuous learning.

Simple Reporting Scheme

A formal incident report should contain the following information, at a minimum:

1. Problem Definition
   What
   When
   Where
   Significance
      Loss
      Frequency
      Safety Issue
2. Summary Statement
3. Corrective Actions and Associated Causes
4. Responsible Person and Completion Date
5. Cost Information or ROI
6. Contact Person
7. Report Date

Keep the report simple. As Albert Einstein reportedly said, "If you can't say it simply, you probably don't understand it." I used to get so many long-winded reports on my desk that I used the manage-by-reputation method to review them. I would read the subject, the summary (if provided), and who wrote the report. If it was written by someone I knew and trusted to do a good job, I often just signed it off. There wasn't enough time in the day to adequately read all reports, let alone gain an appreciative understanding of them. When you consider my strategy (not uncommon today) and the fact that only 30% of the workforce are effective problem solvers, it is no wonder why problems kept happening. The report should be limited to one page and have a Realitychart attached. A typical example of an incident report is shown below, and Figure 5.1 shows the accompanying Realitychart.

Incident Report Form

Purpose:	Prevent recurrence of stated problem; not place blame
What:	Potential electrical contact
When:	10/28/02, during elevator maintenance
Where:	Top of elevator car, West Buchanon Bldg
Significance:	
Safety:	Potentially serious injury or fatality; exposure to high voltage
Environmental:	NA
Production:	No impact
Maintenance:	Repair time 16 hrs; $3,000 capital, total costs $3,950
Frequency:	First time

Summary Statement (See attached Apollo Cause and Effect Chart):

The potential electrical contact was caused by activation of electrical circuits and the proximity of an electrician's hands near energized circuits. His hands were near the circuits because he was performing annual maintenance. The circuit activated because the elevator call button was pushed and an incomplete circuit lockout existed. The call button was pushed because an employee needed to get to the third floor and ignored the "work in progress" signs.

The need to get to the third floor was caused by the need to deliver mail. The signs were ignored because employee believed they were not applicable to him because in his earlier use he was allowed and encouraged to do so by the maintenance electrician. The circuit lockout was incomplete because the lockout program had not been reviewed or upgraded for five years.

Incident Costs		Corrective Action Costs		Return on Investment
Investigation	100 hrs ($5000)	Replacement	N/A	N/A: Safety
Property Loss	N/A	Repair	$3,950	
Lost Sales	N/A	Testing	$1,300	
Production Loss	N/A	Design	$3,400	
Cleanup Costs	N/A	Other	N/A	
Other	N/A*	Total	$8,650	
Total	$5000			

Priority	Cause	Corrective Actions	Person Responsible	Completion Date
2	Lockout Incomplete	Disable Call Buttons before Maintenaince	Robin Strauss	02/13/93
3	Procedure not reviewed	Revise procedure to include current codes and standards and establish annual review cycle by users.	Ed Davis	02/08/93

Contact: Joel Nesham X7819   Date: 11/07/02

*Estimated loss for each OSHA recordable is $25,000; this was a near miss.

The information contained in the report is more important than the form. The report could contain more information than given here, but these items have proven effective in providing enough information to communicate the event and ensure that effective problem solving will occur. You should always create your own form to meet your needs. Do not encourage restrictive thinking by including endless checklists and specific questions such as, "Was the hazardous condition recognized?" These only serve to limit thinking and foster favorite solutions that will ensure a repeat event.

The summary statement is simply verbiage that reflects the core set of causes in the attached Realitychart. The purpose of the summary statement is more to meet traditional expectations than provide any real value. Since the Realitychart provides everything one needs to understand the causes and effectiveness of the solutions, the summary statement is somewhat redundant. However, in its defense, I have found that managers are unwilling to accept such a radical change as to provide only a Realitychart, so I suggest the summary statement be included in the interest of harmony.

Corrective Actions Tracking Program

Using a master list, each corrective action should be documented in a log. The log should include the responsible person(s), completion date, and a brief description of the required action.

The Corrective Actions Tracking Log should be updated frequently and have the highest visibility in the organization. The facility manager or similar authority should review the log weekly or more often if necessary. If a corrective action is not completed on time, an explanation must be provided and a new date assigned. Failure to maintain discipline on this list will be seen as a lack of commitment by management and the entire program will fail.

Corrective actions should be agreed to and signed off by those who have the authority and responsibility to implement them, no one else.

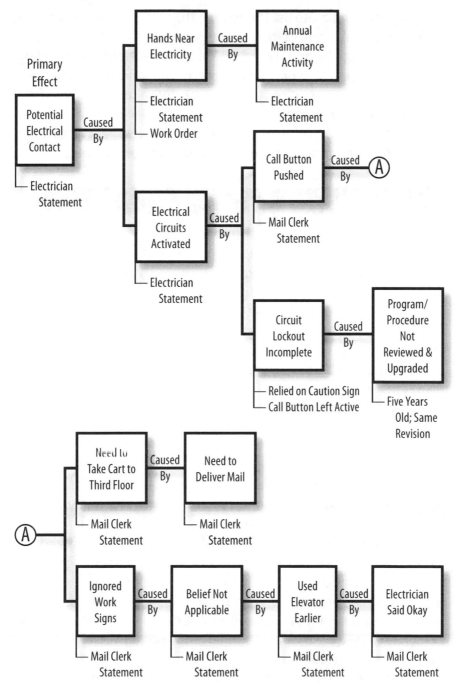

Figure 5.1. Example Completed Apollo Cause and Effect Chart

Create a separate list for actions that call for review, analysis, or investigation. Long-term projects or "nice-to-do" tasks should be kept separate from the master Corrective Actions Tracking Log. The master log should only include those specific items that result from a formal investigation, that is, corrective actions derived from an event significant enough to require a formal investigation and meet the three criteria for an effective solution (prevent recurrence, be within our control, and meet our goals and objectives).

Trending Causes

Today, trending causes and problems is used to help us determine where best to put our resources to realize maximum benefit. Historically, trending has required categorization, but with the use of RealityCharting® and word search capabilities, the need for categorization diminishes. RealityCharting® has a "Find" feature that allows you to search for specific causes or specific causal elements so you can search your past .rca files to identify common repeatable causes which can help identify systemic causes/problems. While this is not the most efficient process at this time, it can be utilized. We are working on providing a feature that will allow you to search any number of completed Realitycharts (called .rca files) at the same time.

If you are still using the old method of categorizing problems or causes and then creating a Pareto chart or other trending information you need to remember that everyone categorizes differently. In many event reports, you may find a set of cause categories on a form that the individual is asked to check. These categories are then tracked and trended and compared. The National Safety Council has been gathering safety data on incidents this way since its inception. What they fail to understand is that because each person has a different perception of the world, they have a different categorization scheme. What one person understands to be a personnel error another person will see as a hardware or procedural failure.

For years I ran an exercise in my training courses where I would ask the students to categorize a list of thirty items into people, procedure, or hardware. We would then compare each item. To everyone's amazement, people would categorize many things differently. Some items are easily agreed upon, but others are not. An interesting thing about this exercise was the incredulity expressed by the students. Because of our fundamental belief that there is a single reality and everyone can see it, it comes as a shock when we see an example of different realities. As an example, one item was patience. Some categorized this as "people" because it is a characteristic of people. Others saw this as a "procedure" because it is a practiced strategy.

The lesson in this is that if you are going to trend causes by using categories, funnel all categorization through one or two like-minded people. This is not a question of what the "right" category is, but one of ensuring consistency. All databases that rely on checklist input from different people are subject to this same discrepancy, and therefore most databases provide bogus information.

While trending causes or problems is an essential element of conventional quality programs today, I submit that it is an ineffective methodology. The accepted purpose of trending problems or causes is the belief that we can make a first cut in the vast problems we have and only focus on the top 20% that cause 80% of the losses. While a valid statistical approach, it is based on the assumption that repeat events are the norm. If you are having repeat events, it either means you have an ineffective problem-solving program or the events are not important enough to warrant your attention.

Organizations that have implemented the Apollo method do not have repeat events, so the need for tracking and trending causes becomes a moot point. If the problems are not significant enough for us to prevent, we should not waste time tracking and trending them.

The key here is to work on all significant problems as defined by your threshold criteria. As the total number of problems becomes smaller because they do not recur, lower your threshold criteria to work on less significant problems. This is what continuous improvement is all about.

Bear in mind that until such time as you have fully implemented the Apollo problem-solving program, you will need to work off your existing trending program. After you work the problems off the current list, all you will need to do is track the total number of events exceeding the threshold criteria. The total number will provide a gauge to help you decide when to lower your threshold criteria.

Implementing Solutions

So then, an effective problem-solving program is one that includes all the elements discussed in this chapter. But, it is more than just following a few steps. By providing simple tools, the Apollo process makes the program work. An honest dedication to continuous improvement is now possible because the Realitychart provides a simple tool to accommodate differences in perception. A questioning environment is now encouraged. Indeed, it is embraced by those who once could not deal with dissension. The common reality created by the Realitychart enables individuals and organizations to more easily follow the dictum of continuous improvement. RealityCharting® makes sharing realities and documenting them easier, resulting in better solutions.

The Apollo problem-solving tools are simple, but as we saw in Chapter 1, they are in conflict with what we normally do. To overcome our natural tendencies to tell stories and communicate categorically, people need to be shown how ineffective current problem solving is and how effective it can be. Effective training must include using real-world examples and comparing current problem-solving methods with the Apollo process.

By using the Apollo problem-solving process and establishing some threshold criteria that dictate when to perform formal incident investigations, we can break out of the useless policy of trending repeat events and focus on prevention instead.

The key to making an effective problem-solving program work is to have a dedicated champion for every business unit. The dedicated champion must be experienced, affable, and respected by most people in the organization. The champion should report at the highest level in the business unit and have the ability to cross all organizational boundaries. The champion is someone who is not only well versed in the Apollo methods, but who also teaches as he/she promotes the concepts and uses RealityCharting® to communicate all problem solving matters. The champion is often the one who is called on to facilitate the investigation of major events. As such, the champion must have special skills in group facilitation. These skills are discussed in the next chapter.

6

Facilitating Groups

Working together to accomplish great things will always be a part of the human experience, and success depends on our individual courage to confront group consensus.

Effective problem solving can be accomplished individually or in a group. The Apollo method will work in either situation, but it is most powerful within a group. A group without a leader is a mob, and a problem-solving team without a facilitator is just another group. Therefore, I strongly suggest that you use the skills that a facilitator will bring to this effort.

Facilitating the problem-solving process using the Apollo method has the same elements as we've discussed:

1. *Gather information.*
2. *Define the problem.*
3. *Create a Realitychart.*
4. *Identify effective solutions.*

This chapter reiterates what we have learned so far, but it puts it into the perspective of facilitating a team. It also discusses the many common traps we encounter when trying to communicate and solve problems together and ways to work around them.

Do not confuse group facilitation, a recognized field unto itself, with facilitating the creation of a Realitychart. While facilitating skills are very helpful in creating a good cause and effect chart, the guidelines in this chapter are more specific to facilitating the Apollo problem-solving process.

A good facilitator does not have to know anything about the problem. It is often a major benefit for the facilitator to know nothing about the problem but to know the Apollo problem-solving method.

Facilitating the Apollo problem-solving method is a process of gathering information, defining the problem, creating a Realitychart, and finding creative effective solutions while practicing appreciative understanding.

Actually facilitating the problem-solving process can be significantly enhanced if the facilitator has good people skills. This chapter is designed to provide some guidance in this area and addresses many of the problems encountered during facilitation. We will also discuss several ineffective human strategies and how to overcome them. The final section provides some Q&A that I hope are helpful.

Facilitation Guidelines

Problem solving is a way of thinking and cannot be relegated to a defined rule set. Therefore, it is with great trepidation that I write the following facilitation guidelines. As tempting as it might be, please do not use this information to create a procedure. To do so is to get rule-based thinking confused with cause and effect thinking. These are only guidelines to be used to get the novice started. As you gain confidence and experience, you will find what works for you.

The facilitator can be the key to an effective problem analysis. The purpose of the facilitator is to ease and promote the problem-solving process. The problem-solving process generally follows a sequence as listed below, with the caveats that gathering information occurs throughout the process and problem definition can change at anytime, even as late as identifying solutions. The sequence of facilitation is as follows:

1. Gather information.
2. Define the problem.
3. Create a Realitychart.
4. Identify solutions

I've listed some guidelines to help facilitators with each of these steps. Because a key aspect of gathering information involves interviewing, that subject is covered in detail in the next section.

Gather Information

While listed as step 1, gathering information is a continual process starting with finding out everything you can about an event and continuing until you verify that the solution meets the three criteria. Gathering information is not an individual task. Everyone in the organization should be trained and understand the need for causes and evidence. If they have not been trained, the facilitator should always give a short explanation of the Apollo process at the beginning of the first team meeting.

Appreciative understanding plays a big role in every part of effective problem solving, but it plays a crucial role when gathering information. Any parochial judgments or biases used to filter or eliminate information at this stage may prove very damaging to your success later on—do not discard anything at this time.

Time is the essence of success when gathering information. Evidence preservation policies should be established, known by everyone, and used assiduously. In manufacturing facilities, an overzealous focus on production often gets in the way of good sense. If we do not take the time to gather evidence on significant problems, chances are they may not be solved.

Tenacity and doggedness are key watchwords in any investigation. Consider the following story of the payoffs it can bring. In 1926 one of the largest producers of tool steel in the United States provided the US auto industry with most of its critical components. About this time the number of auto accidents began to rise and drivers were being killed when steering knuckles began to fail. As we learned in Chapter 3 we tend to believe that our problems are caused by others and the automakers were no exception. They immediately blamed the steel company for providing inferior steel.

The steel companies checked every possible cause at their end and could find nothing. The steel tested correctly and nothing changed during shipment to the automakers. When the automakers gave the steel company an ultimatum to correct this unknown problem an engineer from the steel company was sent to Detroit to observe the automakers use of the steel. All processes were checked and nothing was discovered to suggest any problems. The heating temperatures where checked and were found to be exact, but the problem persisted. Being tenacious by nature this engineer decided to have the temperature gauges checked. He found that all the gauges were manufactured by the same company and indeed they were all found to be faulty. The end result was that while the steel was being made into steering knuckles it was overheated by several hundred degrees higher than it should have been. The problem was solved and steering knuckles did not present a problem again—until recently. In addition to the need to be tenacious, an added lesson here is to validate your inferred evidence when things don't seem right.

Preserve evidence by securing the environment, the people, and the process controls. If dealing with broken hardware, do not touch the broken parts and get them to your material specialist as soon as possible. Limit access to the area or equipment, find out who was involved, and what they know about the event. If process controls such as procedures are involved, identify them and their role in the event. It is often helpful to have a "go-bag" on hand to help in the information-gathering process. A typical investigator's go-bag consists of the following items (make your own list):

- *digital camera*
- *paper and pencils*
- *interview guidelines*
- *evidence preservation checklist*
- *other guidelines*
- *grid paper for mapping*
- *measuring tape*
- *flashlight*
- *labels, tags, and duct tape*
- *steel ruler*

- *feeler gauges*
- *marking pens*
- *sealable plastic bags*
- *small tape recorder for your notes*
- *magnifying glass*
- *magnet*
- *rags*
- *sample bottle(s)*
- *"why" questions—a questioning attitude*
- *mirror.*

Guidance in gathering evidence can be obtained from a reference list called the evidence preservation checklist. A typical industry checklist follows, but you should create your own:

- *Preserve the condition and location of hardware*
  ____ Equipment
  ____ Tools
  ____ Materials (removed or installed)
- *Obtain and preserve documentation*
  ____ Regulations and standards
  ____ Procedures
  ____ Work instructions
  ____ Design drawings
  ____ Operator logs
  ____ Equipment logs
  ____ Process strip charts
  ____ Work requests
  ____ Maintenance records
  ____ Surveillance records
  ____ Quality records
  ____ Work schedules
  ____ Computer printouts
- *Document evidence*
  ____ Photos, sketches, drawings, and maps
- *Collect input*
  ____ Personnel statements
  ____ Interviews or peer review reports.

Based on the initial information and problem definition, determine who will be involved in the problem-solving team. This may change as the need for expert advice is realized. Only invite people who will contribute to the effort. Caution: Limit the number of people in the team to fewer than eight. Four or five is optimal for most events. With a very complicated problem, you may want to create multiple teams to address special areas. These teams should report to the main team by providing their portion of the Realitychart. Always maintain a master Realitychart that everyone can look at anytime during the investigation.

Develop a sequence of events or timeline before you start to create a Realitychart. The sequence of events will provide an initial set of action causes that can be used to help you develop the initial chart.

As you will discover, RealityCharting® has a Wizard that helps you through the entire problem solving process and it should be used by the experienced facilitator. If you are not an experienced facilitator, you should use the following guidelines until you gain the necessary confidence.

Define the Problem

After the initial information gathering, the team should come together with a flip chart or other vertical board to write on. Open the meeting with your expectations and set the following ground rules as applicable:

1. Everyone should strive to appreciatively understand all points of view.
2. Complaining, comparing, or competing is not allowed.
3. The purpose is to fix the problem, not to blame.
4. Everyone is here to contribute.
5. This is an open dialog; judging or stating conclusions are not allowed until the solutions phase.
6. We are not trying to find the right answer; we are going to find the best solutions.
7. We will not talk about solutions until after we create a Realitychart.

8. The best solutions must meet three criteria: prevent recurrence, be within your control, and meet your goals and objectives. Explain as necessary.
9. Be patient with the process; explain it briefly if some people have not been trained.
10. Do not hold side conversations.
11. We are looking for causes and their supporting evidence.
12. Everything is open to discussion, but the facilitator reserves the right to direct the discussion to follow evidence-based causes.
13. Assumptions are encouraged, and they will be labeled with a question mark until we can find supporting evidence.

Unless you are an experienced facilitator and user of RealityCharting®, you should avoid using computers to facilitate the initial cause and effect chart. Use a flip chart or white board to document and spur discussion.

Begin to define the problem by asking the team members to identify the primary effect as they see it. Write each one down.

Listen for cause statements and write them down. Record every notable cause statement on a sticky. Place them on a vertical surface (a wall, whiteboard, or flip chart). (You may delegate the writing to another team member or ask those who suggest them to write it on a sticky and give it to you for placement.) Remember don't judge whether these are primary effects or not, just write down everything. People are providing their perspective and each cause is part of the puzzle.

When no further cause come, ask the group which cause they think is the primary effect and begin putting hem in order from present to past. This will get you to a primary effect that most can agree on.

After you have a general agreement on the primary effect, finish the problem statement by writing the When, Where, and Significance so everyone can see. Remember to provide specific information about safety, cost, and frequency.

Do not proceed until every member agrees with the written problem statements. If you cannot get concurrence, remind dissenters that this can be changed anytime and ask if you can move on.

If there is more than one primary effect, write out a problem statement for each and then proceed to ask "why" of one primary effect at a time.

Create a Realitychart

Starting with the primary effect, begin asking "why?" or asking "caused by?" until you no longer get answers. Many answers will already be in front of you from the problem definition stage; use them.

Encourage an open dialogue. No one judges; if anyone does, remind them of the ground rules. It is not unusual to feel you have lost control at this stage. This is quite normal and can last twenty to thirty minutes. Each individual reality is pouring out and it is usually productive as long as everyone follows the ground rules. This is very similar to working on a jigsaw puzzle; things are not very clear until you get some pieces to fit together. It will come together. After experiencing this a few times, you gain confidence and recognize the out-of-control feeling as normal. Listen carefully and write down every cause you hear without regard for where it fits into the puzzle; that will come later as you go through the Square One Loop.

Remember, causes are noun-verb phrases. Listen for them. To keep everyone interested, validate their ideas by writing their causes on a sticky or ask them to write their causes down and put it on the board. If the cause is valuable, it will fit; if it is not, it will fall away and everyone will see why. Do not waste time at this point trying to judge or evaluate the value of each offered cause.

Minimize discussion during this phase by asking "why" immediately after placing the cause sticky on the vertical surface. This important point keeps people focused and keeps you moving down a productive path. Anything you can do to keep moving prevents storytelling and gets you to a common reality much sooner. Minimizing the drudgery of an investigation makes people want to do this again for other problems. Avoid getting bogged down in endless analysis and storytelling. The difference between a trained team and an untrained team can mean hours in problem resolution time. The well-trained team will come to the meeting with causes and evidence while the untrained team will come with stories and opinions. Everyone should receive some training in the Apollo problem-solving process. The time saved and the quality of the solutions pays for the training many times over.

Go back to the primary effect (Square One) and start through the cause chains again.

Look for causes in actions and conditions. Caution: Do not let this become an obsession. The reason you are looking for actions and conditions is that the cause and effect principle dictates they are there; and the more causes you can find, the better your solutions.

> **Hint:** If you have an action cause and can't find a condition you can often create a condition by taking the noun in the action cause and adding the word "exists" to it. For example, if the action cause is "Ignored Work Signs" a conditional cause is "Work Signs Exists." If you have a conditional cause and are having difficulty finding an action cause look for words that end in –ed. This does not always produce a verb (action), but will help most of the time. For example if your effect is "Potential Electrical Contact" and a conditional cause is "Hands Near Electricity" the action cause could be "Electrical Circuits Activated," as opposed to "Electrical Circuits Active."

Ideas and causes are usually coming so fast on the first pass it is better to keep the momentum going than to slow down the thought process by labeling causes. As you go through the second and subsequent loops, look for the needed action or condition and baby steps. If you can't find the needed action or condition, don't worry about it.

Go to your point of ignorance. Repeat the Square One Loop as many times as needed to get question marks or make a decision to stop at the end of each cause chain.

If you do not want to go further, then stop and identify solutions. Normally you will need to gather more information to find answers to several "why" questions and to find supporting evidence. If this is the case, place question marks at these points and RealityCharting® will create an action items list for you. Assign responsibilities in the action items list and send it to all responsible players. Decide when to reconvene, and dismiss the team until then.

While evidence can be added at anytime as you go through the Realitychart, it is often best to wait until after you have most causes identified. If evidence is not available, develop a plan to obtain the supporting data.

Complete the Realitychart as best you can. Remember it is impossible to know all the causes. Problem significance will help you to know how far to go with baby steps or termination of the cause chains. Time constraints may also limit exploration, but don't muse over this. Your purpose is to find a creative solution that meets your goals and objectives; and if you accomplish that, then you have accomplished what you set out to do. When you get to the solutions stage and you cannot find an effective solution, then work on the chart some more. This is common.

If storytelling erupts, let it go as long as you are getting causes out of the story. As soon as it digresses into who did what at such-and-such a time at such-and-such a place, stop it and get back into the Square One Loop.

Make sure not to stop too soon on each cause path. Before you decide to stop, look at the last cause in each cause path try to ask "why" two more times. If you end up in "la-la land," then you know you went too far. If you get good answers, keep going. The most common tendency is to stop at categorical causes like "Training Less Than Adequate," or "Maintenance Less Than Adequate." Another common stopping point is "Procedures Not Followed." These are categories, not causes, and they must be explained in more detail. Sometimes it helps to ask, "What do you mean by "less than adequate" or "not followed"?

Refrain from discussing solutions and "root causes" while you are constructing the Realitychart.

Once you have a good first cut of causes, create a Realitychart using RealityCharting®. The act of formalizing the cause and effect chart usually causes another iteration. It seems that formalizing causes the mind to re-evaluate what is being presented. In the process, a much clearer picture is usually created. Make sure the team reviews the formal version and iterates it to a higher level of understanding. This often takes many iterations.

Identify Solutions

Once you have decided to stop adding causes to your chart (going from left to right), go to the causes on the right-hand side, and begin to challenge them. Ask, "why is this cause here, or can I remove it or prevent it from acting?"

As you challenge each cause, provide solutions. RealityCharting® provides two easy ways to do this. You can either select Wizard Step 3, Create Solutions and it will take you through each cause on the chart in a structured way, or you can use the Solutions Tool (Light bulb at the top of the work space) which will allow you to add solutions to any cause you click on. Do not be concerned about strict compliance with the solution criteria at this time. RealityCharting® will help you do this later. This is similar to brainstorming in the sense that you should allow unbiased free thought. Get all team members involved in the creative solutions process to build ownership.

> **Hint:** As you gain experience with the Apollo process, you will find the solution criteria are part of your thinking as you consider each cause. While it should not restrict your thinking, it acts as a guide to keep you focused on a solution that prevents recurrence, is within your control, and meets your goals.

Continue challenging the causes. Do not waste time with causes that do not offer good solutions. If no one in the group can think of anything, move on. Normally, this should not take more than twenty minutes as a group activity. If you have time, it is a good idea to let the solutions "cook" for some time. Talk with people outside the group about the proposed solutions or go to the place where the solution will be implemented and try to visualize implementation. This often identifies other problems.

After identifying solutions, check each one against the solution criteria and decide on the best ones. They must meet the criteria, which include not causing other problems and providing good value for your investment.

You may find one solution that will prevent the problem from happening "most of the time." As you affect more causes and add more solutions, you are reducing the probability of a repeat event, but there comes a point of diminishing returns. Only you can decide where that is, based on your goals and objectives.

Some solutions may not prevent the stated problem from recurring but will help create a better environment for success and therefore may warrant implementation.

Be very careful not to stop with your favorite solution or a group consensus that compromises the effectiveness of better solutions.

Conducting Interviews

When gathering information from people, effective interviewing skills can make all the difference. While it is best to get people with the knowledge of the problem into the team that is developing the Realitychart, it is not always practical. In these instances, one-on-one formal interviews may be the best way to get information. While many sources provide information on how to conduct effective interviews, I have accumulated interviewing ideas from thousands of my students over the past ten years. Some of these ideas will work for you, some may not. Use them accordingly.

The purpose of an interview is to gather causes and evidence of a historical event. Everyone will have a different perception, and people will nearly always tell the truth if given the chance. If they believe they will be punished or ridiculed, they may not participate or provide much value to your quest for information. Therefore, assume that everyone is telling their truth; to judge it otherwise during the interview would be a mistake on the interviewer's part.

Never find fault or place blame. It should be a well planned and structured process focused on understanding what the interviewee perceives and should include feelings and evidence-based causes. Interviewing is about listening and empathizing. Listening includes observing nonverbal communication or body language. Many studies have shown that body language provides more than half of our communication, and those who fail to understand this will never be good interviewers.

Look for signs of frustration and ask the interviewee to tell you why they feel frustrated. Don't be afraid to divert from your initial line of questions if you see the opportunity to learn what is in the mind of the interviewee. Current brain studies show gut feelings are very real even though we are usually inadequate at explaining them.[1] Pursue these and other feelings as they relate to the event. Remember, the validity of any cause will be determined by the completion of the Realitychart, so do not attempt to make judgments during the interviewing process.

Prerequisites for Interviewing

The following are some prerequisites for interviewing:

1. The interviewee must have an incentive to participate. An incentive to participate can come from many quarters, but the need to be needed is perhaps the strongest need we humans have, and capitalizing on this is the interviewer's greatest tool. Putting yourself in the position of the student and the interviewee as the teacher will set the stage for an open dialog. Sometimes simply gaining approval or acceptance of their beliefs provides the incentive to participate.

Occasionally people think they don't have anything to offer but have a need to learn what happened. If you think they know something and want to pursue it, their need to learn will provide an incentive. Share what you know and then ask them to fill in any blanks.

There is no incentive if punishment is a consequence. Generally, interviewees are uncomfortable because they are fearful of punishment for themselves, others, or their group. If you have the authority to promise no punishment, then do so. Otherwise try to get them to see the greater value of learning from mistakes.

2. The interviewer must have credibility with the interviewee. If the interviewer doesn't have credibility, an alternative interviewer should be found. An antagonistic interview is worse than none at all

because the negative feedback to the workforce will poison other information sources or future information gathering.

3. The interview must have a clear purpose. Be able to concisely state the purpose of the interview. If we do not understand why we are conducting an interview, it will be the first question the interviewee will ask. Based on my experience, interviews are often seen as a precursor to a hanging, so you need to have a clear understanding of the purpose. Fears can often be allayed by developing a Realitychart of the interviewee's reality because the interviewee sees where you are going with the questions. Be careful not to restrict yourself to the Realitychart unless you are well into the development of it and the purpose of the interview is to finish off a cause chain.

4. Understand the group dynamics prior to the interview. The politics of an organization can get you in trouble quicker than anything. For example, if the interviewee has a vendetta against his or her boss, the information may be very biased. If you do not know the politics or group dynamics of an organization going into an interview, try to find an unbiased person to help you.

5. Be prepared with predefined questions. A formal interview should never be an impromptu activity. Always spend time preparing questions so you have a direction. This does not mean you are restricted to these questions, but it provides initial structure. More often than not, your questions will not be used because the interviewee will take you to places you never dreamed of.

6. Be prepared to listen with an open mind. Listening is the key to effective interviewing. Suspending judgment or maintaining a positive bias while listening is just as important.

7. Dress accordingly. Unfortunately, we judge others by the first impression. One's appearance is an important part of setting the stage for success. Dress up or down

depending on the situation. If you have to go to a dirty place, such as the location of the event, to conduct the interview, wear clothes that can get dirty, not formal or dressy attire.

8. Catch people at a relaxed time. If interviewees are busy performing their job or focused on work, they cannot be expected to have their minds on your questions. Creating a relaxed environment or finding one to conduct the interview can reduce any worries they may have. Meal breaks may be a good time for interviews, although you need to be careful not to impose on someone's time or give the impression that the company won't take its time to investigate the problem.

9. Always meet on the interviewee's turf or at a neutral location. A neutral location is not the conference room next to the boss's office.

10. Sit on the same side of the table with the interviewee if possible. This will help create a relaxed environment, not one of interrogation. Also, it puts people "on the same side" in more than one way.

11. Schedule the interview. If you have an impatient interviewee, schedule the interview first and obtain approval from the supervisor if needed.

Starting the Interview

After an introduction and small talk to create a comfortable environment, explain the purpose of the interview and ask the interviewee if they are comfortable with sharing information. If not, then jump right into why they don't feel comfortable. Perhaps you have another problem or a cause somewhere in your current problem.

Communicate the common goal of preventing problem recurrence and meeting customer needs. Give the interviewee background on what you know about the event and what you hope to learn from this interview. Let interviewees know their names will not be used in any reports and make sure you follow through with this promise. There is no valid reason for including names in a report. Use titles or positions if necessary and be as general as you can be and still convey the message.

Start the questioning with open-ended questions. Open-ended questions are any questions that cannot be answered with a "yes" or "no." The best opening question is, "Please help me understand what happened?"

Avoid presenting yourself as a "know-it-all." Remember, you are the student and they are the experts. Don't be afraid to let them know you are fallible or don't know what is going on. If you did, you wouldn't be asking for their help. You must believe this to play it honestly.

Aspects of the Interview

Go slow at first. Let the interviewee "warm up." Keep on track with prepared questions, but encourage all relevant discussion. Use expressions such as "Go on," "What does that mean to you?" or "Can you explain that further?" to continue a train of thought they may be struggling with. Sometimes it is appropriate to ask for feelings. "How did that make you feel?" can be a valid question if dealing with an emotional part of the event. Cause and effect is not limited to pure logic and reason. "Feeling Upset" is as valid a cause as "Leg Broken."

Always be honest in your dealings with others, but go out of your way to be this way in an interview. If you don't know something, say "I don't know." Sometimes the discussion leads to asking interviewees if they have any ideas how to find the answer. "I wonder who does know about that?" "Do you know anyone who might help us?" is a possible line of questions. Use follow-up questions to help focus the interview on a specific cause path, such as, "And that would mean?" or "So what does that tell us about the causes?" "I don't understand, could you please elaborate?" is another good question to keep the thought process going. This helps build the interviewee's confidence that they are doing the right thing and are needed.

Maintain eye contact as much as possible to pick up on body language. If you are not skilled in reading body language, start learning or ask someone else to perform the interview. Eye contact is acceptable in most western cultures but may be rude or even dangerous in some cultures.

Be enthusiastic about answers if they provide new insights. Do not be phony in this endeavor, but let interviewees know they are being helpful. Restate information by paraphrasing. This helps understanding. A good paraphrase should include a summary of the essential words (key nouns and verbs, not the modifiers), the emotional level from which the statements are made, and what value the interviewee places on them.

Avoid criticizing, complaining, or comparing and ask the interviewees to do the same if you catch them doing it. When responding to questions, use "I" not "we," "them," and "us" as this sets up a perception of different camps and can cause the interviewee to choose sides.

Take good notes or ask someone else to take notes while you do the interview. If you have a note taker, make sure you explain this function at the beginning of the interview.

Using "dead air" or an extended pause after a question often forces the interviewee to think more deeply about the question. We do not like dead air and hence feel a need to fill it with something. What may seem like a long time often is not when things are quiet. This takes practice and doesn't always work, so use it according to your skills and the need to get the conversation going.

Closing the Interview

When the questions subside and there appears to be nothing more to learn, it is time to close the interview. A good question at the end of the interview is, "What would you do differently?" or "If you had a million dollars to spend on this problem, what would you do?"

Review your notes with the interviewee. If legible, show your notes to interviewees and ask them if they see anything you left out or they want to add. The act of showing them your notes builds trust. You may even want to tell them you are going to do this at the beginning of the interview. Give a brief review of how helpful the interviewee has been (if true) and restate and write down any action items or commitments to get more information.

Explain what will be done with the information and promise to get back to them when the report is finished. Never pass up a chance to thank them and ask if there are any more questions before you leave. Ask if they can think of anyone else who might have information about this event or who could shed some light on the subject in general. Give interviewees your phone numbers and ask them to call you anytime they think of additional information. A good closing question is to ask if there was anything they expected you to ask that you didn't. Often people go into an interview with preconceived ideas and prepared answers. When you never get to their prepared answers, they may feel you didn't do your job but not speak up for fear of embarrassing you.

Dealing with Personnel Performance Issues

Interviewing someone who has been involved in less than stellar performance can often be very difficult for both the interviewee and the interviewer. While never easy, I have found that having internalized the cause and effect principle helps a great deal. By knowing that causes are infinite in nature and knowing that people do not purposefully attract negative attention by setting out to screw up, you can bust through the "I screwed up" barrier.

Starting with the effect of "screwed up" or "human error," ask "why." As the individual who had the personnel performance problem reflects on the causes, they will focus on their actions. The person may say they pushed the wrong button, or said the wrong thing, or moved the wrong way. Each of these causes is an action cause so try to identify the corresponding conditional causes that existed in time before their action.

Remember the cause and effect principle teaches us that every action has at least one conditional cause that existed in time before the action set the chain in motion to cause the undesirable effect. Finding these conditional causes often results in a big "aha" for everyone. More often than not, the individual has been set up to fail by the conditions of the task. When someone says, "I just screwed up," it should be a red flag for the interviewer. People do not come to work with the intent of making a mistake. Sure, some people do not pay attention or are incapable of learning, but it is incumbent on the facilitator to determine this with evidence-based causes, even if the individual believes he or she screwed up. This requires getting down to the causes between the causes and looking for human-based conditions.

If the action was "pushed the wrong button," you may want to determine what conditions were presented to the person. Was there enough light? Was the labeling correct? Was there adequate training and knowledge, etc? The most difficult conditions to find and confront are the ones that lie within the human mind, such as "tired," or "confused."

I have found the following list provides a good source of ideas for questions to be asked when dealing with personnel performance issues:[2]

- *Too much information for the mind to comprehend*
- *Boring task*
- *Not proficient in task*
- *Unaware of action causes*
- *Lack of confidence, people, procedures, or hardware*
- *Success in past experiences*
- *Weariness or fatigue*
- *Confusion*
- *Reactive response*
- *Memory lapse*
- *Fear of failure*
- *Priorities misaligned*
- *Spatially misoriented*
- *Inattention to detail*
- *Rigid mindset*
- *Myopic view of situation*
- *Scheduling pressure to complete task*
- *Lack of specific knowledge necessary to complete task*
- *Habit*
- *Inappropriate assumptions*
- *Used shortcuts*
- *Did not understand instructions*
- *Job performance standards not defined*
- *Disbelief in sensory input*
- *Used favorite indication instead of diverse input*
- *Indifferent attitude*
- *Illness*
- *Righteousness*
- *Inability to focus on task.*

This list can be used to find possible cause paths for the lack of performance. For example, you may want to ask if the individual was fearful of failing or if they were sick or tired at the time of the action. If none of these apply or lead you to a clearer understanding of the causes, then perhaps it was inattentiveness or a failure to learn. Don't forget to continue asking "why" as you break through an emotional barrier with one of these possible causes.

A classic example of breaking through can be shown when pursuing the causes of an operator action or improper action. When asked why the operator made a mistake, the answer often comes back as "did not follow procedure." The investigation stops, and some favorite solution such as "rewrite the procedure" is offered and accepted. The problem here is stopping too soon. Break through this cause by continuing to ask "why" several more times until you get to "la-la-land" or the fuzzy zone. There are many reasons why people do not follow procedures and stopping at "did not follow procedure" will result in guaranteed recurrence.

Common Traps

Along the way to effective solutions, we encounter many traps. These are primarily ineffective human strategies that get in the way of the cause and effect principle. These traps are sometimes unique to the individual and sometimes apply to the team or group. In addition to the causes of ineffective problem solving discussed in Chapter 1, the following is a list of common traps I have observed: consensus, groupthink, experts, parochial mindset, programmatic barrier, denial, and time as a cause.

I will discuss each one in some detail so you can recognize them more readily. I will also offer some guidance on how to deal with each one.

Consensus

The belief that the majority rules is so ingrained in our democratic belief system that consensus takes on the appearance of being a fundamental principle. While a very useful strategy, seeking consensus can be very detrimental. Most people are followers and want to be lead. Tell them a good story they can connect with and they will follow along. The consensus trap follows this logic: "I am not really sure what is going on, but the collective knowledge of the group certainly could not be wrong, so I will go along."

The more people who follow, the more the consensus effect grows. Good leaders know this and capitalize on this herd mentality. While we use consensus to make decisions, we need to understand that consensus is an agreement to take a risk together, nothing more. If we want to minimize the risk, we need to base the decision to agree on evidence-based causal relationships, not innuendo and storytelling. An effective solution occurs because we understand the causal relationships, not because the consensus voted on it. Always use a Realitychart in your decision process.

Groupthink

Groupthink[3] is a term coined by the noted research psychologist Irving Janis (1918–1990) from Yale University and a professor emeritus at the University of California, Berkeley which described the systematic errors made by groups when making collective decisions. Groupthink is the condition of relinquishing our individuality for the perceived common good of the group. In fact, this perceived rightness by the group is a form of consensus that will doom the success of the group. Groupthink is subtle to the group unless they know what to look for. It is found in any group of people of any size working or playing together. It can be found in a married couple or an organized religion. It is a fundamental human condition and has proven very detrimental in our history.

The disaster at Pearl Harbor in 1941 was caused by the firm belief of Admiral Kimmel and his small staff that the Japanese would never attack them. The strength of this group belief, called groupthink, was unreasonable to a fault. Even when the bombs started falling, they thought it was a drill and couldn't understand who had authorized a drill on Sunday morning. Groupthink is characterized by many symptoms:

- *A belief that the group can do no wrong.*
- *A belief that the group has a higher authority than any individual inside or outside the group.*

- *Rationalization to justify a position established by the group, regardless of what other factors may be present.*
- *A strong sense of them and us.*
- *An atmosphere to conform. Anyone dissenting is ridiculed or put in a bad light to encourage conformity. Consensus holds the highest priority.*
- *Individual censorship based on the belief by individuals that they couldn't possibly be as smart as the entire group.*
- *The belief in group unity. Without ever calling for a vote, it is assumed that everyone in the group agrees to the same position. Sometimes this occurs even when the position goes unstated. The need for unity is so strong, potential conflicts are avoided or denied.*
- *Individuals speaking for the whole group.*

To prevent groupthink from getting started in your group or team, foster open discussion on any subject. Use a Realitychart to create a common reality based on evidence-based causes not storytelling or opinions of the strongest personality. Ask everyone to play devil's advocate and ask an outsider to review your work, if possible. Honestly address their comments. Avoid sharing conclusions outside the group discussion.

If you recognize the symptoms of groupthink as listed above, do the following:

1. Share what symptom you sense with the group. Let everyone know you think the group is falling into the trap of groupthink. If you get people who disagree strongly with your observations, then the group is probably engaging in groupthink.
2. Ask to be educated; play dumb with probing questions that bring a different perspective to the table.
3. Let team members know it is okay not to know, then work on developing a plan to find answers.
4. Use the Realitychart as the basis for your common reality. Remember to use evidence and always go to your point of ignorance and find out if someone outside the group can answer your questions.

5. Encourage outside points of view and take action to bring them into the group discussion.
6. Challenge all statements that are made by an individual speaking for the group, such as, "I think we can all agree."

Groupthink is a strong human trait and difficult to recognize because it feels so natural to belong to a winning team that we don't want to upset the appearance of successful decisions.

Experts

Experts are essential to effective problem solving, but they should not be given any more credibility than the next person with evidence-based causes. By definition, the expert is a narrowly focused person who knows a great deal about his or her subject matter, but they do not know everything. Experts have a tendency to be right-minded. If they present themselves as opinionated and having the correct solution, beware. Everyone has to play by the same rules when using the Realitychart. If the expert provides a cause, they must also provide evidence. If they have none, do not get into an argument, but put a question mark in the evidence box and move on.

As a young engineer I was taught to respond to clients' questions with the statement; "It is my engineering judgment that this is true." This seems like a rather shallow response to technical questions, but it worked. What amazed me was how readily this was accepted by the client, and how many engineers believe it to be an acceptable practice. The non-engineer has no comeback for such a statement and is left to accept it. It is not a question of whether the statement is valid, it is a question of making the statement without any evidence to support it. As a young engineer I didn't know any better, and it was common practice in my design engineering organization. I did not seek to defraud or mislead, rather to express an opinion and add some credibility to it with the rather baseless "engineering judgment" statement.

As I grew older and began to realize the misuse of this term not only in my profession but in all experts from laborer to physician, I stopped using it. Part of being human is wanting to speak authoritatively. If people let us get away with it, we use this strategy with great skill. When you encounter this strategy, avoid a contest of wills and the spewing of excrement from large barnyard animals and just ask for the evidence to put on your Realitychart. Often the opinion is based in fact, but the expert has not been challenged for so long they have forgotten. Ask for references, examples, or details of past experiences.

Parochial Mindset

Parochial mindset or provincial thinking is yet another human condition that limits effective problem solving. My travels throughout the world have convinced me that it exists everywhere. It is a significant barrier to effective solutions because it drives us right to our favorite solutions. It is the common belief within a group that if no one in the group knows the answer to a question, then there is no answer to be had anywhere.

The next time you are working on a problem with others, step back and watch the discourse. As the questions and answers unfold, eventually a question will go unanswered. That is, someone will ask a question and the air is still with silence. After a short pause, because we don't like dead air, someone will change the subject or ask a completely different question. At this point there occurs an unstated but conscious agreement by all players that there is no answer and any pursuit of one has no value. The cause chain is stopped and never followed up unless someone in the group understands and implements the rules of the Apollo process.

This is a most interesting observation to me because it is a totally irrational act on the part of very rational people. I have observed this in almost every group since I first discovered it. When asked why they stopped asking "why," the team members acknowledge they stopped but seem perplexed that I am asking about it. They seem to believe it is obvious why they stopped. When I ask them if they think someone outside the group might have an answer, they will acknowledge the possibility but will not pursue it unless pushed.

Always go outside your group for answers to the unanswered "why" questions. It is incredibly arrogant to think that you or your group are the only ones on the planet who know what's going on.

Even if you are working on a specialized problem within a specialized industry, there is usually someone else who may know about these causal relationships and be able to provide some insights. Go to your local university; they love to work on real-world events. Stopping too soon is a common reason for ineffective solutions.

Programmatic Barrier

Another cause of stopping too soon on a cause path is the programmatic barrier. This is similar to the parochial mindset but has different origins. When following a cause chain, the programmatic barrier occurs as we reach a point where the answer to the next "why" question will result in questioning some organizational program. This seems to be caused by the fear of discussing an institutionalized program. To attack an established program will require much effort and probably not yield any changes, so we stop. Sometimes the solutions associated with the last cause are general. For example, we may find that someone stopped at "people not adequately trained." Knowing that the training program is inviolate may cause the team to offer a solution of retraining without regard for the causes of ineffective training. To break through this barrier, always go to your collective point of ignorance on every cause path.

Denial

Denial is the strongest human attribute we have. Its roots are in groovenation, but it is manifested in many ways. The need to maintain our own reality is sometimes stronger than the need to learn new cause and effect relationships. Sometimes perceptions intrude upon our "realities" and cause major conflicts. I recently came across an example of this in the responses to a National Geographic article on lions feeding at night. One letter to the editor read as follows:

"I found the photographs very unsettling. They captured the victim animals at their most private and vulnerable moments— those of terror and death. I am outraged at your assumption that I want to see these struggles."

This person openly stated her wish to deny her perception because it conflicted with her "reality." Furthermore, she is "outraged" that someone else could see a different reality and want to share it.

A different reality was presented in another letter to the editor regarding the same story.

"My daughter, age four years nine months, looked over my shoulder as I was reading my August issue. She was so interested in the pictures of the lions that I had to read all the captions to her with minor deciphering of difficult words. She now understands that a night in the life of a lion is not exactly as it is for Simba in the movie, The Lion King."

This parent not only enjoyed the sight of her reality, but shared it with her child.

While watching a television news magazine, I saw yet another incredible display of denial and how opinions become fact. The Chief of Public Health for the State of Kentucky enlightened the viewers with this logic. "If tobacco sickness were real, we would know by now because we have been growing tobacco for over two hundred years." He said this with the full knowledge that hospitals and other medical facilities treat hundreds of people each year for tobacco sickness. Tobacco sickness is a common problem with people who work cutting and storing tobacco plants. The freshly cut tobacco secretes nicotine, a highly toxic alkaloid that is absorbed through the skin. It causes fainting, weakness, and sometimes death. But, according to some people, dying from tobacco sickness is not really a problem because it has been happening for over two hundred years.

When you observe someone denying what is in front of them, ask them to provide sensory evidence. Barring this, ask them to explain the causal relationships that support their views. This will usually help them overcome the misconception, but don't be surprised if they cannot offer an explanation. Denial is an incredibly strong aspect of the human condition.

Time As a Cause

Don't use time as a cause. Listen closely to our excuses. The cause is often given as time. We hear examples of this logic in daily conversation:

- *The reason my car looks so bad is because it is old.*

- *I couldn't finish my project because I ran out of time.*
- *I was late to work because time got away from me.*
- *We would have won the game if only there was more time.*

Things happen in time not because of time. The car does not look old because of time, nor is it worn out because of time. It is worn out because of use and the second law of thermodynamics—entropy, the natural law that dictates everything in the universe is trying to obtain it lowest energy state. There are many natural processes, such as friction and radiation that cause wear, and they happen in time not because of it.

Dealing with Group Interaction

As you go through the Square One Loop in a group setting, you will find four general types of interaction.

The Proverbial Storyteller

The storyteller will want to take you back to the scene of the problem and tell you all the people involved or give you a history lesson on why things are done the way they are. While this is often interesting and even informative, don't let them take control of the process. Listen carefully to what they are saying, and the first time you hear the answer to your "why" question, write it down, stop the story regardless of where it is, and ask why this cause happened.

Repeat this interruption process until you have mined all their causes. This will do more to shorten your meeting time than anything else you can do. Usually these storytellers begin to see what you are doing and realize that you only want causes and evidence. Because you are making progress and writing down what they have told you, they do not get upset with all the interruptions. Typically, they know you are trying to facilitate the process and will follow your lead, provided you are respectful and cordial. As you progress, they begin to see a better picture than was in their own head.

The Analytic

The analytic is interested in "why" questions but is more interested in sharing the correct answer. Since they have typically analyzed the problem in great detail, their primary purpose is to make sure you understand the correctness of their ideas. More often than not, they have a very narrow perspective of the situation and have left out many other cause paths. They will even tell you why their perspective is the only possible one. Remind them of the infinite set of causes and interact with them just like you would the storyteller. Ask for a cause; as soon as you get it, interrupt them and ask "why" to that cause or that set of causes.

The analytics are more likely to become upset with you, so be patient with them. You don't want to turn off the information supply. The best method I have found to deal with this is to write down every cause they give you. This validates their worth and they are more willing to let go. If they are really getting off track, ask them if you can let that cause path go for awhile and work on the other paths that may seem more productive.

> **Caution:** If you are a storyteller or analytic, you should not be a facilitator unless you have had some facilitator training.

The Nonparticipant

The causes of nonparticipation can be numerous, but they often lie in the "don't know nothing" category or the "don't want to play this game" category. For those who honestly don't know anything about the problem, ask them for insights they may have after you have a pretty good set of causes. The "dumb questions" are often the best. If they say they don't understand it, ask them to tell you why. If the Realitychart doesn't make sense to someone, it is missing something. Use these people as your sounding board and honestly listen to them and make sure you understand why they don't see something. Also enlist their help in the solutions phase.

For the person who does not want to play, the cause could be fear of embarrassment or fear of implicating themselves or others. In either case, let them know there is no wrong answer. They can say anything that makes sense to them; and if it fits into the Realitychart, it will be incorporated. For those fearful of being blamed, let them know the purpose of this process is to find a solution that prevents recurrence, not to place blame or punish. Be careful not to give the assurance of no punishment unless you have that authority. Sometimes managers will usurp the investigator and your credibility is forever destroyed. This is often a tight balancing act because in about 1% of the situations, punishment may be the best solution. Try to identify the possibility of punishment before getting into the details of problem analysis. If it is possible, do not grant amnesty in these cases but continue to develop the Realitychart.

The Participant

The participant is eager to learn and understand what happened. This eagerness is sometimes slow in coming because of painful experiences in previous group problem solving, but it will come in time. The true participant is usually quick to pick up on the basic rules of this process and the importance of causes and evidence. They begin to realize the facilitator is more interested in "why," and the "who" question is never asked. With a consistently honest approach in asking "why," the participants gain confidence and open up as more causes are understood. When someone knows the answer to a sincerely asked question, it is hard for them not to share what they know. It is especially hard if they can see how much clearer the picture will be when they add their knowledge to the common reality being created. People fundamentally want to help others, but they must be assured they will not suffer the pain of embarrassment. This can be accomplished by letting everyone know there is no such thing as a right or wrong answer in this process; there are only causes and evidence.

Facilitation Guidelines: Some Q&A

The following guidelines respond to commonly asked questions. They are intended to provide a quick reference if you get into trouble while facilitating or want a quick review before starting.

How do I maintain meeting focus?

Stay in the Square One Loop; focus on "why" and how you know (evidence). Minimize storytelling and over-analysis by forcing the team to focus on causes. Explain the process to newcomers if needed.

How do I handle team outliers?

For those that already have the right answer, ask them to let the process work for awhile. Remind them that you will get to the solutions only after you know all the causes. If this doesn't work, ask them how they know their answer is the best one. As they explain, listen for causes not yet shown and add them to the Realitychart. Often we know a good solution but don't know why. Our unconscious mind has already figured it out. We call this "gut feel" and it can be very effective if we understand the causal set behind it so others can appreciate it and be assured it is based in causes.

For the defensive person, let them know the purpose of this process is to fix the problem, not to place the blame. Be very careful here; if you do not have the authority to grant amnesty, don't offer it. Never ask "Who did this?"

For the boisterous or assertive person, ask them to hold their comments, and remind them you are looking for causes and evidence, not stories. Explain the difference if needed.

Remember, the need to be needed is the strongest human need; use it to your advantage, such as with the shy person. "Please help me to understand what happened here."

For excuse givers, ask them to define the problem as they see it. Listen carefully and write down the causes they give you. Insert them into the Realitychart and then ask for evidence. If they have no evidence, put a question mark under the cause and move on.

How do I stimulate discussion?

Everyone has an opinion—ask for it.

Everyone needs to be needed. Ask people to help you figure this out using questions such as, "Please help me to understand what happened."

Be dumb like a fox; ask simple probing questions.

Use provocation; make an absurd statement or challenge conventional wisdom.

Use small talk to get people relaxed, then ask for feelings or perceptions.

How do I prevent manipulation of the Apollo process?

If someone on the team is trying to manipulate the causes to exclusively show their reality, make sure you follow the first three steps of the Apollo process without exception. This is often obvious because the chart stops too soon and has few branches.

To force a broader perspective, look for actions and conditions at every node.

Always demand evidence.

Challenge the obvious and conventional wisdom—it is often biased and incomplete.

Challenge or test the belief that the solution will prevent recurrence.

Go outside the group for a separate review.

Be on the alert for groupthink.

How do I avoid embarrassing the participants?

Follow the fundamental rules of chart development.

Establish an open learning environment from the very beginning.

Never ask, "who did it?"

Avoid any judgmental statements.

Write down every stated cause.

Let the team members know that in the Apollo process, there is no such thing as right and wrong, only causes and evidence.

Is evidence that critical?

Evidence is one of the most important elements of the Apollo process. If you fail to use it, you may be setting yourself up to fail. Having said that, it is less important than getting all perspectives to fit on the chart.

Try to find sensory evidence and if you do not have it, use inferred evidence. Try to identify two or more ways to document evidence.

How do I ensure precise cause statements?

Use noun-verb statements.
Try to limit the number of words to less than four.
Avoid prepositions in the cause statement if you can.

How do I write clear corrective actions?

Always use specific corrective actions and include those responsible for implementing them and the completion date.
Avoid using "re-" words, such as "retrain."
Never use study or analyze; if you do, you are not done.
If you cannot connect a solution to one of the causes on the Realitychart, either your solution or your chart is incomplete.

How can I use cause categorization?

Cause categorization should be avoided except as guidance when you cannot get any answers to the "why" questions. Look for causes in people, procedures, hardware, and the environment by asking what role each category played. This will lead you to more specific "why"-type questions.

What are the qualities of a good finished product?

Assiduous adherence to the first three phases of the Apollo process will ensure an effective solution. A well-defined problem statement, a complete Realitychart with evidence, and solutions that attack one or more causes on the chart and meet the three solution criteria constitute the essential elements of a good report.

What do I do if no causes come?

Look for causes in actions and conditions.
Look for causes in categories.
Look for differences and when you find them, ask "why."
Use other problem-solving tools, but always come back to the Realitychart.

How do I resolve a stalemate discussion?

If you have a group of headstrong people who do not value appreciative understanding, ask one team member to create a strawman Realitychart based on how they see the problem. When completed, this strawman chart will be used by all team members to criticize and tear apart, so don't give this task to the right-minded egotist, who knows the right answer. Since it is always easier to criticize than create, the strawman Realitychart moves the group from a creative consensus mode, which is not working, to a completely critical mode, which is building on a new common reality.

How do I overcome storytelling?

Use the Square One Loop.

Let the storyteller go until you hear some causes. As you hear the causes, write them down. As long as causes are coming from the story, let them go; but if the story digresses to discussion of people, places, and things as a function of time, stop them. Put the causes on a Realitychart and ask the team to help you put them in order using "why" and "caused by." When you get all of them placed, pick a cause at the end of a chain, and ask the storyteller, "Why this cause?" Continue this process until you run out of causes.

Is fostering goodwill worth it?

Fostering goodwill may seem like an extra step not worth taking, but problem solving is a continuous part of doing business and anything to promote it is worthwhile. Always send a copy of the final report to everyone who helped in the problem-solving process and thank them for their help. Give special thanks to extraordinary help. Celebrate all major successes by letting everyone know the value added by the solutions. If you had a problem that has occurred ten times over the past five years and it cost $10,000 for each failure, calculate the savings this is going to create and publish it to the broadest audience you can.

Facilitating Groups

The Apollo problem-solving process is the most effective when used in a group. Facilitating a group to effective solutions can be challenging, but it is very rewarding when you arrive at corrective actions without conflict and argument.

The best advice I can give for anyone engaged in group facilitation of the Apollo process is to have faith in the process. It will work if you just follow the first three steps: define the problem, create a Realitychart, and identify effective solutions. Within each success, we always have small failures, such as someone who doesn't want to participate or the proverbial storyteller who won't quit talking, but if you are reflective, you will learn from each incident and get better with time.

With each new incident, you will be presented with the vast collection of human perspectives. Wisdom will lead you to better understand the notion that there is no such thing as one right answer, but because there is an infinite set of causes, you can always find an effective solution. The Apollo problem-solving method has many virtues and they are summarized in the next chapter.

References

1. Carter, Rita 1999 *Mapping The Mind*, University of California Press, London, UK
2. Human Performance Enhancement System Coordinator Manual, INPO 86-016 (Rev. 3), Atlanta, Georgia, 1986.
3. Groupthink: Psychological Studies of Policy Decisions and Fiascoes by Irving L. Janis

7

A New Way of Thinking

A mind once expanded never regains the
same shape.
— Adapted from Oliver Wendell Holmes

*Event-based problem solving is simple once we understand the cause
and effect principle. It is my hope that with this book more people will
begin to appreciate how much more effective they can be at problem
solving, both in their business lives and personal lives. This chapter
summarizes the Apollo problem-solving method and the benefits that
can be realized. The Apollo problem-solving method has the following
attributes:*

- *Can be used by everyone.*
- *Offers a structured approach.*
- *Applies to all event-based problems.*
- *Does not require checklists or forms.*
- *Minimizes storytelling.*
- *Creates a common reality.*
- *Encourages a questioning attitude.*
- *Provides a platform for creative solutions.*

*Effective solutions for everyday problems are guaranteed every time if
you choose to implement these tools.*

When it comes to event-based problems, effective problem solving has long eluded us. It doesn't matter what industry, what company, or what country, the average problem-solving effectiveness for most organizations is about 30%. Repeat events are so common we develop trending programs to measure them and fail to see the contradiction this presents. With formal training, the information provided in this book has improved problem-solving effectiveness to around 95% in some companies.

The Apollo problem-solving method is effective because it is principle-based and works naturally with all points of view. It works in conjunction with all perspectives to allow a common reality to emerge from the diversity of each stakeholder. As we have seen, the methodology is simple in structure and form. It can be used by anyone on all event-based problems without the use of checklists or forms. It counteracts the ineffective human strategies of storytelling and groovenation by creating an evidence-based common reality of cause and effect relationships. By appreciatively understanding all perspectives, the methodology encourages a questioning attitude and serves as a platform for creative solutions anchored in fact, not fantasy or delusion.

By providing structure and form to problem solving, we can now begin to teach problem solving as a subject unto itself. By breaking out of the old paradigm that problem solving is inherent to the subject matter, we can begin to teach people how to think and communicate in a way that provides effective solutions to event-based problems every time. Simply by knowing that there is an infinite set of causes and that every effect has at least two causes we can break out of the linear thinking that has prevented effective problem solving since the beginning of time.

The Apollo problem-solving method is to effective solutions what mathematics is to accounting and engineering. Before we had numbers, humans had very limited accomplishments. Without numbers we could not measure nor engage in serious trade. But a formal numbering system did not happen overnight. Ascribed to Pythagoras, mathematics was not considered a subject until about 500 B.C., and it wasn't until the renaissance of the seventeenth century that mathematics fully blossomed and eventually lead us to the industrial revolution.

Like the evolution of mathematics, the struggle to find a better way to communicate is slow. We have not developed a rule set by which to effectively and consistently solve event-based problems until now. By recognizing the disparity between our linear language and the nonlinear physical world we live in, we see the elegance of the Apollo cause and effect chart. We finally have a tool set that allows all stakeholders to visualize the causes of any event-based problem.

Our linear thinking has misguided us into the narrow-minded thinking of a root cause for every problem. This thinking is born of our language, or perhaps it is the language that shapes our thinking. Either way, when we limit ourselves to think only in terms of A caused B and B caused C, we limit our ability to address the complex issues we must face in our rapidly changing world. This is not to say we should stop telling stories and forget about simple linear cause chains. This is not going to happen and it should not happen. Telling stories is one of the greatest things about being human, and short cause chains often occur in the cause and effect charts.

Expressing branched cause paths like those presented in Chapter 2 is too difficult for our modern languages. By producing a Realitychart, we are providing a visual dialog that enables all stakeholders to learn together such that they arrive at a common solution.

People walk away from an Apollo problem-solving session with the gratification that their perspective was included and the causes make sense. Everyone is confident the solution will work because they can see the causal connections between solution and primary effect. Intuition and gut feelings are even represented if the group concurs with the value. There is a realization that while they did not arrive at their original conclusion, things have to be done differently.

If someone did not participate in the original construction of the chart, it facilitates future visual dialog to accommodate new perspectives. This learning process is enabled by the visual dialog and is a better motivator for improvement than implementing what somebody else dictates.

A Simple Structured Approach

As with a wheel or anything of great value, it is the simplicity that provides the greatest worth. We can take the man out of the cave, but it has been virtually impossible to take the cave out of the man. While we humans have come a long way toward improving our lifestyle, we still have the same brains we had when we lived in the caves. We are very simple-minded creatures living in a very complex world, and by working together we have accomplished things no individual could ever hope to do.

When we come together to accomplish things as a group, we tap the power of synergy, but not without some difficulty. As we learned in this book, as much as we would like to think we are created the same and want to be the same, we are as diverse and unique as each snowflake that falls. Finding a way to overcome the discord this causes and the problem of a linear language in a nonlinear world while accommodating the simple human mind has been a challenge for the ages. I believe this challenge has been met with Apollo Root Cause Analysis.

Effective Solutions for Everyday Problems Every Time

While the Apollo Root Cause analysis problem solving method is highly effective, it requires a new way of thinking for most people. It usually takes two days of exercise intensive training and two or three incident investigations to become an effective user. In nearly 20 years of teaching these methods history shows that students often revert back to old ineffective habits after leaving the classroom. Changing our lifetime strategies of storytelling and categorization is very difficult without help. Over the years, we tried refresher training and taught managers to be better reviewers, but these solutions had limited success at institutionalizing better problem solving. RealityCharting® software was created specifically to help solve this problem by first providing a simple user friendly charting tool and backing it up with rules that follow the Apollo Root Cause Analysis process. RealityCharting® will provide an accurate cause and effect chart every time. It provides the structure needed by the new users to keep them from reverting to old habits.

Whether you are a professional incident investigator, a manager, or a worker, RealityCharting® will help you understand your problem better than you have ever been able to understand it before. As a result of this understanding, you are able to find more effective solutions and effectively communicate the value of those solutions to others. Because the Apollo method and RealityCharting® do not allow storytelling, the normal arguing and politics associated with problem solving is avoided. Clear evidence-based causal relationships are very hard to argue with and the Apollo process encourages diverse ideas and viewpoints such that the best solutions can be found together, as a team. The bottom line is: Using RealityCharting® in concert with a comprehensive training program results in higher quality solutions in less time.

If you are interested in learning more about ARCA and RealityCharting® please visit us at www.realitycharting.com

Appendix

Comparison of Common Root Cause Analysis Tools and Methods

Having studied and worked with most of the Root Cause Analysis (RCA) tools and methods in use today, I am often asked to compare them with RealityCharting®. To satisfy this request, I will compare the various RCA tools and methods to what we have learned about effective problem solving in this book.

This Appendix provides a short description and evaluation of the current and most common RCA tools and methods used in businesses throughout the world. If you want a more in-depth discussion of them, Reference 1 provides one of the better comparisons, but it was written before Apollo RCA was created. Tools are included along with methods because the tools are often touted and used as a full blown root cause analysis, when in fact there is a clear distinction between them and methods.

Comparison Criteria

If we are to evaluate the many so-called root cause analysis methods and tools, we need a standard to which they can be compared. It is generally agreed that the purpose of root cause analysis is to find effective solutions to our problems such that they do not recur. Accordingly, an effective root cause analysis process should provide a clear understanding of exactly how the proposed solutions meet this goal.

To provide this assurance, an effective process should meet the following six criteria.[1]

1. Clearly defines the problem and its significance to the problem owners.
2. Clearly delineates the known causal relationships that combined to cause the problem.[2]
3. Clearly establishes causal relationships between the root cause(s) and the defined problem.
4. Clearly presents the evidence used to support the existence of identified causes.
5. Clearly explains how the solutions will prevent recurrence of the defined problem.
6. Clearly documents criteria 1 through 5 in a final RCA report so others can easily follow the logic of the analysis.[2]

Various RCA Methods and Tools in Use Today

As you will discover in this analysis, there is a clear distinction between an RCA method and a tool. A tool is distinguished by its limited use, while a method may involve many steps and processes and has wide usage.

Events and Causal Factors Charting: (Method) A complicated process that first identifies a sequence of events and aligns them with the conditions that caused them. These events and respective conditions are aligned in a time-line. Events and conditions that have evidence are shown in solid lines but evidence is not listed; all others are shown in dashed lines. After this representation of the problem is complete, an assessment is made by "walking" the chart and asking

---

1. It should be noted that there is value in all of the tools discussed herein, as they all help us better understand our world. The question in this discussion is which should you use to be the most effective problem-solver in your world.
2. It is important to understand the difference between connecting causes based on similar attributes such as a taxonomy of causal factors and connecting them based on how nature actually works. More on this later.

if the problem would be different if the events or conditions were changed. This leads to causal factors that would then be evaluated using a tree diagram (discussed below).

Change Analysis: (Tool) A six-step process that describes the event or problem; then describes the same situation without the problem, compares the two situations and writes down all the differences; analyzes the differences and identifies the consequences of the differences. The results of the change analysis is the cause of the change and will frequently be tied to the passage of time and, therefore, easily fit into an Events and Causal Factors Chart, showing when and what existed before, during and after the change. Change analysis is nearly always used in conjunction with an RCA method to provide a specific cause, not necessarily a root cause.

Barrier Analysis: (Tool) An incident analysis that identifies barriers used to protect a target from harm and analyzes the event to see if the barriers held, failed, or were compromised in some way by tracing the path of the threat from the harmful action to the target. A simple example is a knife in a sheath. The knife is the threat, the sheath is the barrier, and the target is a human. If the sheath somehow fails and a human is injured, the barrier analysis would seek to find out why the barrier failed. The cause of this failure is then identified as the root cause.

Tree Diagrams: (Method) This type of root cause analysis is very common and goes by many names[1] such as Ishikawa Fishbone Diagram, Management Oversight and Risk Tree Analysis (MORT), Human Performance Evaluations System (HPES), and many other commercial brands. These methods use a predefined list of causal factors arranged like a fault tree. (See Figure A.1.)

They are sometimes called "Pre-Defined Fault Trees." The American Society for Quality (ASQ) and others often call these categorical methods "Cause-and-Effect Diagrams." All categorization methods use the same basic logic. The premise is that every problem has causes that lie within a pre-defined set of categories. Ishikawa uses Manpower, Methods, Machinery' and Environment as the top-level categories. Each of these categories has sub-categories and sub-sub-categories. For example, within the category of Manpower, we may find Management Systems; within Management Systems we may find Training; and within Training we may find Training Less Than Adequate; and so on. These methods ask you to focus on one of the categories such as People and in reviewing what you know of your event choose some causal factors from the pre-defined list provided. Each categorical method has its own list of causal factors. After reviewing the list for each category, you are asked to vote on which causal factors most likely caused your problem. After some discussion, the most likely ones are voted on and called root causes. Solutions are then applied to these "root causes" to prevent recurrence. Each commercial brand has a different definition of root cause, but it is generally a cause that you are going to attach a solution

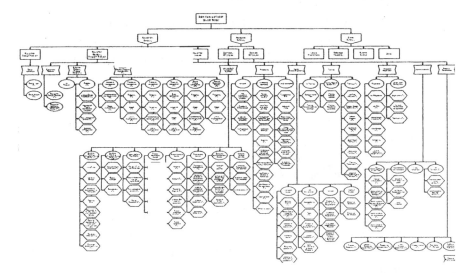

Figure A.1. One Branch of a Tree Diagram

to that prevents recurrence. Some of these methods refer to themselves as "Expert Systems" and also provide pre-defined solutions for your problems.

Why-Why Chart: (Method) One of many brainstorming methods also known as the "Five-Whys" method. This is the most simplistic root cause analysis process and involves repeatedly asking "why" at least five times or until you can no longer answer the question. Five is an arbitrary figure. The theory is that after asking "why" five times you will probably arrive at the root cause. The root cause has been identified when asking "why" doesn't provide any more useful information. This method produces a linear set of causal relationships and uses the experience of the problem owner to determine the root cause and corresponding solutions.

Pareto Analysis: (Tool) A statistical approach to problem solving that uses a database of problems to identify the number of pre-defined causal factors that have occurred in your business or system. It is based on the Pareto principle, also known as the 80-20 rule, which presumes that 80% of your problems are caused by 20% of the causes. It is intended to direct resources towards the most common causes. Often misused as an RCA method, Pareto analysis is best used as a tool for determining where you should start your analysis.

Storytelling Method: (Method) This is not really a root cause analysis *method* but is often passed off as one, so it is included for completeness. It is the single most common incident investigation method and is used by nearly every business and government entity. It typically uses predefined forms that include problem definition, a description of the event, who made a mistake, and what is going to be done to prevent recurrence. There is often a short list of root causes to choose from so a Pareto chart can be created to show where most problems come from.

Fault Tree Analysis: (Method) Fault Tree Analysis (FTA) is a quantitative causal diagram used to identify possible failures in a system. It is a common engineering tool used in the design stages of a project and works well to identify possible causal relationships. It requires the use of specific data regarding known failure rates of components. Causal relationships can be identified with "and" and "or" relationships or various combinations thereof. FTA does not function well as a root cause analysis method, but is often used to support an RCA. More later.

Failure Modes and Effect Analysis: (Tool) Failure Modes and Effects Analysis (FMEA) is similar to fault tree analysis in that it is primarily used in the design of engineered systems rather than root cause analysis. Like the name implies, it identifies a component, subjectively lists all the possible failures (modes) that could happen, and then makes an assessment of the consequences (effect) of each failure. Sometimes a relative score is given to how critical the failure mode is to the operability of the system or component. This is called FMECA, where C stands for Criticality.

Realitycharting*: (Method) A simple causal process whereby one asks why of a defined problem, answers with at least two causes in the form of an action and condition, then asks why of each answer and continues asking why of each stated cause until there are no more answers. At that time, a search for the unknown is launched and the process is repeated several times until a complete cause and effect chart, called a Realitychart, is created showing all the known causes and their inter-relationships. Every cause on the chart has evidence to support its existence or a "?" is used to reflect an unknown and thus a risk. All causes are then examined to find a way to change them with a solution that is within your control, prevents recurrence, and meets your goals and objectives. The result is clear causal connections between your

Method/Tool	Type	Defines problem	Defines all causal relationships	Provides a causal path to root causes	Delineates evidence	Explains how solutions prevent recurrence	Easy to follow report	Score
Events & Causal Factors	Method	yes	limited	no	no	no	no	1.5
Change Analysis	Tool	yes	no	no	no	no	no	1
Barrier Analysis	Tool	yes	no	no	no	no	no	1
Tree Diagrams	Method	yes	no	no	no	no	no	1
Why-Why Chart	Method	yes	no	yes	no	no	no	2
Pareto	Tool	yes	no	no	no	no	no	1
Storytelling	Method	Limited	no	no	no	no	no	0.5
Fault Tree	Method	yes	yes	yes	no	yes	no	4
FMEA	Tool	yes	no	Limited	no	Limited	no	2
RealityCharting®	**Method**	**yes**	**yes**	**yes**	**yes**	**yes**	**yes**	**6**

Figure A.2. Comparison of Selected RCA Methods and Tools

solutions and the defined problem. Because all stakeholders can see these causal relationships in the Realitychart, buy-in of the solutions is readily attained.

RCA Methods and Tools Compared

Many purveyors of Root Cause Analysis state the process is so complicated that you should use several of them for each problem or select them based on which type of problem you are experiencing. In researching the various proponents of this approach I find that the reason some people think root cause analysis is so complicated is they don't understand the cause and effect principle. To quote Albert Einstein, "If you can't say it simply, you probably don't understand it."

Using the comparison criteria we established earlier, Figure A.2. provides a summary of how each method or tool meets the criteria. One point is scored for each criteria that is met. "Limited" is scored as 0.5 points.

While the comparison in Figure A.2 serves to show how poorly these conventional tools and methods provide effective solutions, it does not tell the whole story, as explained below.

Events and Causal Factor Charting can provide the time-line to help discover the action causes, but is generally inefficient and ineffective because it mixes storytelling with conditional causes, thus it produces complicated relationships rather than clarity.

Change Analysis is a very good tool to help determine specific causes or causal elements, but it does not provide a clear understanding of the causal relationships of a given event. Unfortunately, many people who use this method simply ask why the change occurred and fail to complete a comprehensive analysis.

Barrier Analysis can provide an excellent tool for determining where to start your root cause analysis, but it is not a method for finding effective solutions because it does not identify why a barrier failed or was missing. This is beyond the scope of the analysis. To determine root causes, the findings of the barrier analysis must be fed into another process to discover why the barrier failed.

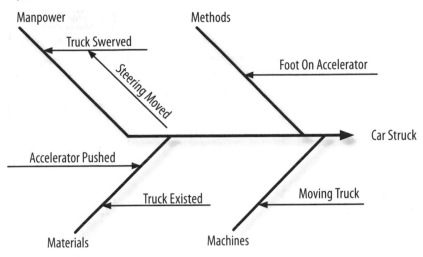

Figure A.3. Fishbone Diagram

Tree Diagrams, a.k.a. Categorization Schemes, are steadily being replaced with RealityCharting® but continue to retain a few followers because they appeal to our sense of order and "push button" type thinking (as discussed in Chapter 1). Based on what you have learned in this book, you can now understand why this is a failed strategy. There are at least 7 major weaknesses in the Tree Diagram model:

1. A Tree Diagram is clearly not a "Cause and Effect Chart" as the proponents of these methods would have us believe. It simply does not show all the causal relationships between the primary effect and the root causes. Consider the following example: Given a simple event, I have arranged the causes according to the rules of a Fishbone Diagram in Figure A.3.

As we can see, the causal relationships are not clear at all. Could it be "Car Struck" was caused by "Foot on Accelerator" and "Truck Swerved" and "Truck Existed" and "Moving Truck?" Certainly these are some causes, but their relationships are not apparent. The diagram was created by looking at the event as I understand it, asking what causes could be classified as Manpower, Methods, Materials, and Machines and then placing those causes on the fishbone according to the categories they belong in—not how they are connected causally. The theory behind these Tree Diagrams is

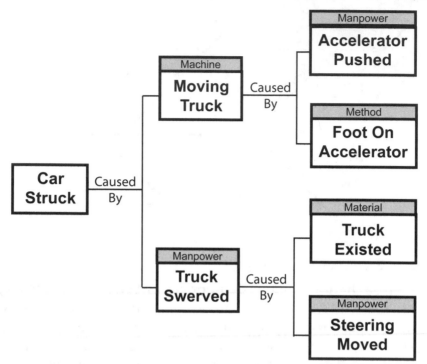

Figure A.4. Realitychart vs. Categorization

that because all events have certain causal factors we can find the root causes by looking for them in the pre-defined set provided. And while it can help jog the mind into certain lines of thinking, it fails to provide a causal understanding of the event.

If we use this same event and create a Realitychart (Figure A.4.) we can clearly see the causal relationships. I have added the categories to the top of each cause to emphasize how knowing the category provides no value whatsoever.

2. No two categorization schemes are the same, nor can they be, because as discussed in Chapter 1, we each have a different way of perceiving the world.[2] Therefore, we have different categorical schemes and hence the reason there are so many different schemes being sold. When asked to categorize a given set of causes it is very difficult to find a consensus in any group. For example, what category does "Pushed Button" fall into? Some will see this as hardware; some will see it as people; and some will see it as procedure.

If you have ever used any of these categorization methods to find a root cause, I suspect you have incurred many a wasted hour debating which is the correct category.

3. The notion that anyone can create a list of causal factors that includes all the possible causes or causal factors of every human event should insult our intelligence. Ask yourself if your behavior can be categorized in a simple list and then ask if it is identical to every other human on the planet. The very fact that a method uses the term causal "factor" should be a heads-up that it does not provide a specific actionable cause but, rather, a broader categorical term representing many possible specific causes. At best, it acts as a checklist of possible causes for a given effect, but it does not provide any casual relationships. Since this error in logic is very contentious with those who use these methods, it begs the question why do these methods seem to work for them. What I have discovered after talking with many people who claim success in using these methods, is that it works in spite of itself by providing some structure for the experienced investigator whose mind provides the actual causal relationships. It is not the methodology that works, but the experience of the investigator who is actually thinking causally. And while these methods seem to work for the experienced investigator, they are still incapable of communicating the reality of causal relationships. This inability to effectively communicate prevents the synergy among stakeholders necessary to fully understand the causes of the event, which is required to get buy-in for the solutions.

4. These models do not provide a means of showing how we know that a cause exists. There is no evidence provided to support the causal factors in the list, so it is not uncommon for causal factors to be included that are politically inspired with no basis in fact. With these

methods, the best storytellers or the boss often get what they want, and the problem repeats. This may help explain why many managers and self-proclaimed leaders like this method.

5.   Categorization schemes restrict thinking by causing the investigator to stop at the categorical cause. Some methods re-enforce this fallacy by providing a "root cause dictionary," implying that it is a well-defined and recognized cause.

6.   Categorization methods perpetuate the root cause myth discussed in Chapter 1, based on the belief it is a root cause we seek and solutions are secondary. Because these methods do not identify complete causal relationships, it is not obvious which causes can be controlled to prevent recurrence; therefore, you are asked to guess and vote on which causal factors are the root causes. Only after root causes are chosen are you asked to identify solutions and without a clear understanding of all causal relationships between the solution and the primary effect, this method works by chance not by design.

7.   As mentioned earlier, some of these methods provide what is called an "expert system" and includes solutions for a given root cause. Expert systems can be quite useful for a very specific system such as a car or production line where 99% of the causal relationships are well known and have a long history of repeatability. To presume that one could provide an expert system applicable to all event-based problems seems to me to be incredibly arrogant. In light of what you now know about the infinite set of causes that governs reality, how could anyone presume to know the causes for all systems, how they interrelate, and what constitutes the best solution for every organization or individual? Beware the salesperson.

The Five-Whys method is inappropriate for any complicated event, but it is actually quite useful when used on minor problems that require nothing more than some basic discussion of the event. Unlike most of the other methods, it identifies causal relationships, but still subscribes to the root cause myth of first finding the root cause and then assigning solutions. It should never be used for formal incident investigations, but is perfectly acceptable for informal discussions of cause. A popular graphical representation of the "Five-Whys" approach is the "Why Staircase," which if used improperly leads to a linear set of causal relationships.

Pareto Analysis uses a failure database to trend the frequency of categorical failures. As discussed in Chapter 5 under Trending Causes, this process is fraught with many landmines, a few of which are discussed below.

1. The accuracy of a Pareto chart is limited by the accuracy of the data used to create it. If you use a failed approach like tree diagrams to determine the causes, the Pareto chart will only reflect causes from the pre-defined list provided.

2. As you learned in this book, the cause and effect principle shows that all causes and effects are part of the same continuum. It many cases, certain causes will be closely linked (i.e. close to each other). For example, the cause "procedures not followed" is frequently caused by "procedures not accurate." In the Pareto analysis, this causal connection is lost. Instead, we see both "procedures not followed" and "procedures not accurate" in those top causes, so we end up working on solving both problems when in reality we may only need to solve the "procedures not accurate" problem. In this example, the incomplete view of reality provided by a Pareto analysis may have caused you to expend more resources than necessary.

3. Pareto analysis can mask larger, more systemic issues. For example, if quality management has transitioned into a state of dysfunction, it can cause symptoms in many different areas, such as poor procedures, inadequate resources, outdated methods, high

failure rates, low morale, etc. Pareto analysis has you capturing all these symptoms of a larger problem as causes, and wasting time solving the symptoms.

Storytelling: Perhaps the most common of all methods is storytelling, also known as the fill-out-a-form method. This method was discussed in more detail in Chapter 1 but is summarized here for consistency. The primary difficulty with this approach is that you are relying completely on the experience and judgment of the report authors in assuring that the recommended solutions connect to the causes of the problems. The precise mapping between the problem and the recommended solutions is not provided.

The primary purpose of this method is to document the investigation and corrective actions. These forms usually do a good job of capturing the what, when, and where of the event, but little or no analysis occurs. Consequently, the corrective actions fail to prevent recurrence 70% to 80% of the time.

With such poor results, you might be wondering why organizations continue to use this method. The answer is twofold. First, most organizations do not measure the effectiveness of their corrective actions, so they don't know they are ineffective. Second, there is a false belief that everyone is a good problem-solver, and all they need to do is document it on a form. For those companies that recognized they are having repeat events, a more detailed form is often created that forces the users to follow a specified line of questions with the belief that an effective solution will emerge.

This is a false promise because the human thinking process cannot be reduced to a form. In our attempt to standardize the thinking process, we restrict our thinking to a predefined set of causes and solutions. The form tacitly signals the user to turn off the mind, fill in the blanks, and check the boxes. Because effective problem solving has been short circuited, the reports are incomplete and the problems keep occurring.

Fault Tree Analysis is not normally used as a root cause analysis method[3], primarily because it does not work well when human actions are inserted as a cause. This is because the wide variance of possible human failure rates prevents accurate results. But it works extremely well at defining engineered systems and can be used to supplement an RCA in the following ways:

1. Finding causes by reviewing the assumptions and design decisions made during the system's original design
2. Determining if certain causal scenarios are probable, and
3. Selecting the appropriate solution(s).

Additional insight into the various RCA methods, and how RCA integrates with quantitative methods such as fault tree analysis can be found in Reference 3.

Failure Modes and Effect Analysis: Failure Modes and Effects Analysis (FMEA) is sometimes used to find the cause of a component failure. Like many of these other tools, it can be used to help you find a causal element within a Realitychart. However, it does not work well on systems or complex problems because it cannot show evidence-based causal relationships beyond the specific failure mode being analyzed.

Realitycharting*: Realitycharting is unlike all other RCA tools and methods. It is the only one that actually provides a graphical representation with evidence of all causes and their inter-relationships. With this clear understanding of your reality, it can easily be communicated to anyone with a full appreciation of how the solutions will prevent the problem from recurring.

Summary

While conventional root cause analysis tools provide some structure to the process of human event problem solving, this review shows how they are significantly limited and often work by chance not by design. The common processes of storytelling and categorization are the product of thousands of years of evolution in our thinking, but it is time to move on. RealityCharting° is becoming the standard for all event analysis because it is the only process that understands and follows the cause and effect principle, thus it is the only process that allows all stakeholders to create a clear and common reality to promote effective solutions every time.

* **A note on terminology:** What we used to call "Apollo Root Cause Analysis" is being replaced by the term "Realitycharting." The end result, a Realitychart, is being requested by name. Informed managers require Realitycharts for all major problems and employees ask to be trained because they see how it can help them be more successful. Realitycharting has become a core competency in many companies because the return on investment is overwhelmingly positive.

References

1. Wilson, Paul, et al., 1993, Root Cause Analysis – A Tool for Total Quality Management, Quality Press, Milwaukee, WI
2. Churchland, Paul M., 1996, The Engine of Reason, the Seat of the Soul, MIT Press, Cambridge, MA
3. Reising, Larry; Portwood, Brett, 2007, Root Cause Analysis and Quantitative Methods - Yin and Yang?; 2007 Paper presented at the International System Safety Conference (a copy of this paper can be found at www.realitycharting.com).

Glossary

The following terms are used in the text of Apollo Root Cause Analysis.

actions: Momentary causes that bring conditions together to cause an effect, sometimes called action causes.

appreciative understanding: Prerequisite mindset for effective problem solving that is characterized by suspending judgment and maintaining a positive attitude. It occurs by accepting all incoming information at face value and allowing the ARCA process to determine value.

ARCA: Apollo Root Cause Analysis. While this term is very common in those industries who have adopted it, it is being replaced with the term Realitycharting because a Realitychart is the end product of the ARCA process.

baby-step causes: Causes between the causes that represent a finer look at causal relationships. All causal relationships have causes between the causes, but we are typically too ignorant to see them.

categorical thinking: The natural process of the mind that orders all knowledge into specifically defined classes. It can present a significant barrier to effective problem solving.

cause: The answer to a "why" question. It should be stated as a noun and verb and it comes in two forms; an action cause and a conditional cause.

causal element: Referred to in Chapter 2 as the elemental causal set, it is the fundamental set of causes that consists of at least one condition, and one action for a given effect.

cause and effect principle: The principle that every effect has causes that follow four characteristics, which are: 1. Cause and effect are the same thing. 2. Causes and effects are part of an infinite continuum of causes. 3. Each effect has at least two causes in the form of actions and conditions. 4. An effect exists only if its causes exist at the same point in time and space.

chunking: The breaking down of a large event into small manageable events.

common reality: A combined reality created from the individual realities of several people and documented by a Realitychart. If an individual's reality does not fit into the common reality, it is likely because the person cannot provide causal evidence or they are seeing a different problem.

common sense: A mythical common feeling of humanity. It is an illusion that causes ineffective problem-solving by assuming that everyone sees the same world.

conditions: Causes that exist over time before an action brings them together to cause an effect, also called conditional causes.

effective problem solving: Identifying causal relationships and controlling one or more of the causes to affect the problem in a way that meets our goals and objectives. A key goal of event-based problems is to prevent recurrence.

effective solution: A solution that prevents problem recurrence.

elemental causal set: The fundamental causal element of all that happens. It is made up of an effect and its immediate causes and represents a single causal relationship. The causes consist of an action and one or more conditions. Causal sets, like causes, cannot exist alone. They are part of a continuum of causes with no beginning or end.

event: An interaction of causes at a particular place and time. The smallest event is an elemental causal set. The largest event has an infinite number of causes.

event-based problems: Problems that center around people, objects, and rules that occur in time and space. These type of problems are distinguished from rule-based problems by having more than one possible solution.

evidence: Data used to conclude something. Evidence comes in different quality levels: sensed by the five senses, inferred, intuited, or emotionally sensed.

fact: A cause supported by evidence. Facts outside of a causal relationship have no value. For example, the fact that the shoe is brown is really a characteristic of the shoe, while calling it a fact may be grammatically correct, it misrepresents the meaning of a "fact."

favorite solution mindset: Our natural tendency to seek a familiar solution to problems based on some categorical assessment. This strategy is ineffective most of the time.

garbage solution strategy: Placing as many problems as possible into a category and then solving the problem categorically, like the way we put all our garbage into a single container and it magically disappears. While the solution appears to work, it causes many other problems.

groovenation: The process of justifying our beliefs. It is physiological in origin and is found in our search to validate our existing realities.

groupthink: The condition of relinquishing our individuality for the perceived common good of a group.

ignorance: Lack of knowledge. A term that humans should embrace as their enemy to be overcome, but commonly scorned because of their egocentric nature. Commonly used in this book to provoke thought.

point of ignorance: The point where we can honestly admit we don't know when repeatedly asking "why."

Only one in twenty people are capable of reaching this point.

primary effect: Any effect of consequence that we want to prevent from occurring.

problem solving: Overcoming a difficulty by implementing a solution.

prototypical truths: Conclusions about the world we live in that are subject to change given enough evidence to support a new conclusion. All our truths are prototypical; some are just more ensconced than others.

root cause: Any cause in the cause continuum that is acted upon by a solution such that the problem does not recur. It is not the root cause we seek, it is effective solutions.

rule-based problems: Problems that do not require people, objects, or time and space and always have a predefined right answer.

single reality: Sometimes called the truth. Most humans believe there is a single reality that everyone can see, but what we fail to understand is that no two humans perceive the world the same. It is physiologically impossible for any two people to possess the same view of the world. With this dilemma, the best we can hope for is a common reality.

solution: An action taken upon a cause to affect a desired condition.

solution killers: Very judgmental statements used to kill a solution idea, for example, "We already tried that once." Used by fearful people to resist change of any kind.

Square One Loop: Following each cause path in an Apollo cause and effect chart until the collective point of ignorance is reached, and then starting over again with the primary effect (square one) and repeating the process. Each time through the "loop" we look for baby steps, cause branches, and evidence.

stakeholder: A term taken from Dr. Stephen R. Covey's Principle-Centered Leadership[1] which describes all people who have a stake in the success or failure of an enterprise or group.

storytelling: Communication describing an event by relating people, places, and things in a linear time frame from past to present.

strategy: An ordering process used by the mind to organize knowledge to solve problems.

system: A causal set that includes one or more feed-back loops.

References

1. Principle-Centered Leadership, Stephen R. Covey, Simon & Schuster, Inc., New York, 1992.

Index

Acknowledgements

This book is the product of many years of study driven by my unwillingness to accept contradictions. I see contradictions where most people do not, so I have studied the sciences, religion, and philosophy, and I am grateful for all those who have gone before me in trying to figure out the notion of causation. Aristotle, Aquinas, Pascal, Newton, and several modern players like Peter M. Senge have provided valuable stepping stones to the message of this book. I am indebted to my students around the world who have asked invaluable questions and challenged me to know more. Without their questioning attitude, I would never have come this far.

I am particularly grateful for the intellectual stimulation of my colleagues, Larry Reising, Chris Eckert, and Brian Hughes. Along with many others who I have encountered along the way, my son, Wesley J. Gano, provided an honest intellectual sounding board necessary to break outside the envelope of conventional wisdom.

I had many new challenges in preparing this book, so I am grateful to my wife, Mary, and my parents for teaching me to greet every new challenge with a positive attitude.

Honest comments are hard to come by when one presumes to write a book, so I am very thankful for the editorial reviews and critical comments provided by those listed above as well as Fred Logan, Mark Hall, Greg Prior, and E. J. Ledet.

Editors always win in the end and for this we writers are thankful. I want to acknowledge my editors, Vicki E. Lee of Lee Scientific Communications and Sue Gano. I want to thank them for their patience, dedication to a smooth-flowing book, and for putting up with my protestations and analytical mindset.

Along the way to understanding the Apollo problem-solving process, I had the greatest difficulty trying to understand the notion of evidence. As hard as I searched, I could not find any good reference material on the subject. Two people were instrumental in helping me come to the understanding in this book. Over the course of four years, Jennifer L. Patterson and

E. J. Ledet both asked the most insightful questions and, in trying to answer their concerns, I figured it out.

A special thanks must go to Joe Glass, formally of Union Carbide Chemicals, who believed in me and gave me a chance to prove the value of this material in the world of industry. With his help and that of many Carbide people like Paul Balmert, the Apollo methods were continually tested and improved until they worked for everybody.

Another special thanks goes to Ted Bennett and Fred Logan of Dow Chemical Company for their trust in the Apollo method. Through their leadership and dedication to continuous improvement, we learned how to institutionalize a new way of thinking into a large global operation which continues today.